Materials, Chemicals and Methods for Dental Applications

Scrivener Publishing
100 Cummings Center, Suite 541J
Beverly, MA 01915-6106

Publishers at Scrivener
Martin Scrivener (martin@scrivenerpublishing.com)
Phillip Carmical (pcarmical@scrivenerpublishing.com)

Materials, Chemicals and Methods for Dental Applications

Johannes Karl Fink

Scrivener Publishing

WILEY

This edition first published 2018 by John Wiley & Sons, Inc., 111 River Street, Hoboken, NJ 07030, USA and Scrivener Publishing LLC, 100 Cummings Center, Suite 541J, Beverly, MA 01915, USA

For more information about Scrivener publications please visit www.scrivenerpublishing.com.

Wiley Global Headquarters
111 River Street, Hoboken, NJ 07030, USA

For details of our global editorial offices, customer services, and more information about Wiley products visit us at www.wiley.com.

Library of Congress Cataloging-in-Publication Data
ISBN 978-1-119-51031-4

Cover images: Pixabay.Com
Cover design by: Russell Richardson

Set in size of 11pt and Minion Pro by Exeter Premedia Services Private Ltd., Chennai, India

10 9 8 7 6 5 4 3 2 1

Contents

Preface

This book focuses on the materials used for dental applications. The text focuses on the basic issues and also the literature of the past decade. The book provides a broad overview of dental materials.

Chemicals that are used for the preparation and fabrication of dental materials are explained. Also, the desired properties of these materials are discussed. The relevance of the chemical, physical, and mechanical properties is elucidated.

Methods for the characterization and classification, as well as clinical studies are reviewed here. In particular, materials for dental crowns, implants, toothpaste compositions, mouth rinses, and also materials for toothbrushes and dental floss are discussed.

For example, in toothpaste compositions several classes of materials are incorporated, such as abrasives, detergents, humectants, thickeners, sweeteners, coloring agents, bad breath reduction agents, flavoring agents, tartar control agents, and others. These chemicals are detailed in the text and the chemical structures for a lot of these chemicals are also shown in the figures.

This text may be of importance for students of dental hygiene, therapy and nursing. Anincrease of knowledge regarding the chemical issues in this field is expected. Also, chemists and other scientists that are interested in these topics could be interested in reading this text to expand their basic knowledge.

How to Use This Book

Utmost care has been taken to present reliable data. Because of the vast variety of material presented here, however, the text cannot be complete in all aspects, and it is recommended that the reader study the original literature for more complete information.

Table 1.1 History of dental polymeric materials (3).

Event	Date
Synthesis and polymerization of methyl methacrylate	1901
Use of poly(methyl methacrylate) (PMMA) as denture base resin (Germany)	1930
First acrylic filling material	1944
Addition of inorganic fillers (non-bonded) to direct filling materials	1951
Investigation of epoxy resins as direct filling materials	1955
Acid-etch technique introduced (Buonocore)	1955
Dimethacrylates (Bis-GMA) and silanized inorganic filler investigated as direct filling material (Bowen)	1958
Bis-GMA composites marketed	1964
Development of polymeric coatings on fillers (Dental Fillings Ltd)	1968
UV-cured dimethacrylate composite resins	1973
Visible-light-cured dimethacrylate composite resins	1977
Introduction of Filtek™ Silorane System to the market	2007

1.2 Properties

1.2.1 Acronyms for Compounds in Dental Compounds

Common acronyms for compounds used in dental formulations are collected in Tables 1.2 and 1.3. Some of these compounds are shown in Figure 1.1. Initiators and inhibitors are shown in Figure 1.2.

1.2.2 Standards in Dentistry

Standards that are used in dentistry are summarized in Table 1.4.

Color stability is an important characteristic of dental materials and it is expected that the test methods in this International Standard ISO 7491:2000 will be referred to in the International Standards specifying such materials (5).

ISO 10477:2004 classifies polymer-based dental crown and bridge materials and also specifies their requirements. In addition, this standard specifies the test methods to be used to determine the compliance with these requirements (6). It is applicable to polymer-based dental crown and bridge materials for laboratory-fabricated

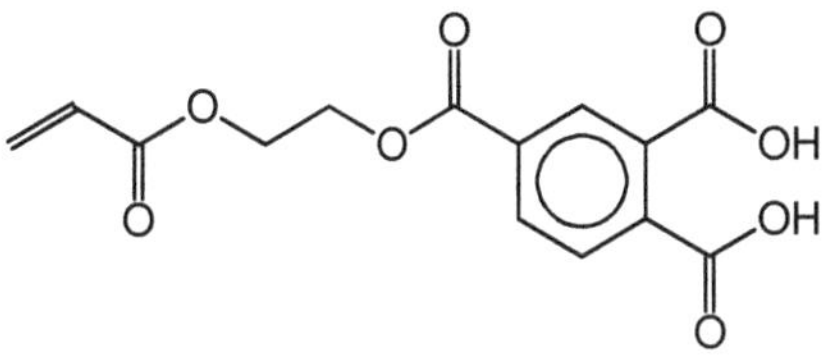

4-Acryloylethyl trimellitic acid

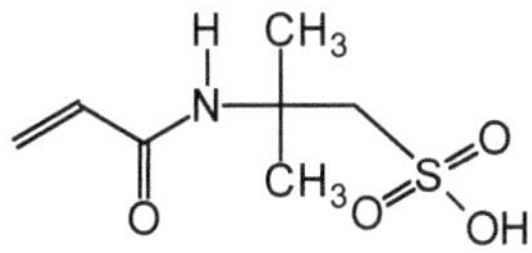

2-Acrylamido-2-methyl-
1-propanesulfonic acid

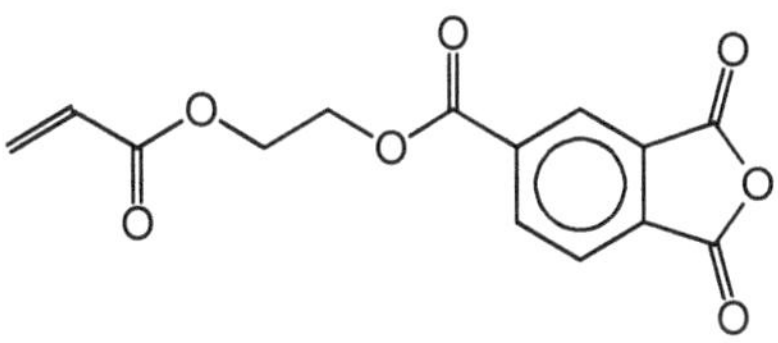

4-Acryloyloxyethyl trimellitate anhydride

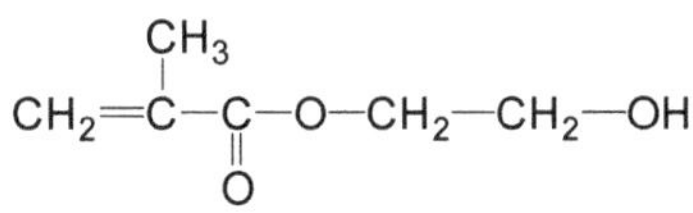

2-Hydroxyethyl methacrylate

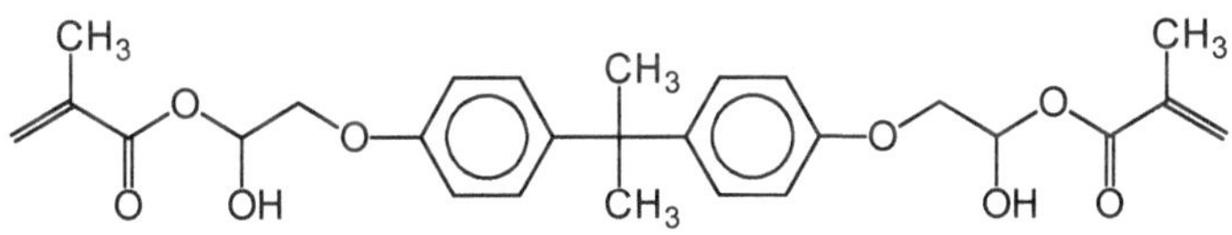

Bisphenol A diglycidyl methacrylate

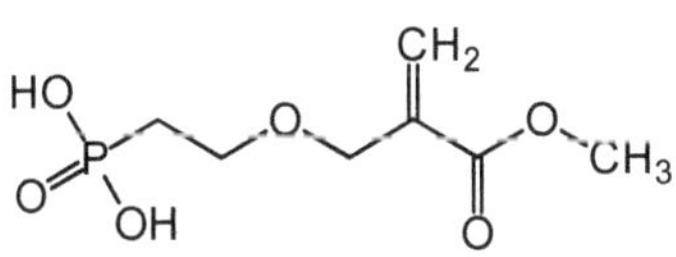

Ethyl 2-[4-(dihydroxyphosphoryl)-2-oxabutyl]acrylate

Figure 1.1 Monomers used in dental formulations.

Table 1.2 Acronyms for monomers (4).

Acronym	Chemical Name
4-AETA	4-Acryloyloxyethyl trimellitate anhydride
4-AET	4-Acryloylethyl trimellitic acid
AMPS	2-Acrylamido-2-methyl-1-propanesulfonic acid
Bis-MEP	Bis[2-(methacryloyloxy)ethyl] phosphate
Bis-EMA	Ethoxylated bisphenol A glycol dimethacrylate
Bis-GMA	Bisphenol A diglycidyl methacrylate
BPDM	Biphenyl dimethacrylate or 4,40-dimethacryloyloxy-ethyloxycarbonylbiphenyl-3,30-dicarboxylic acid
DMAEMA	Dimethylaminoethyl methacrylate
EAEPA	Ethyl 2-[4-(dihydroxyphosphoryl)-2-oxabutyl]acrylate
EGDMA	Ethylene glycol dimethacrylate
GDMA	Glycerol dimethacrylate
GPDM	Glycerol phosphate dimethacrylate
HDDMA	1,6-Hexanediol dimethacrylate
HEMA	2-Hydroxyethyl methacrylate

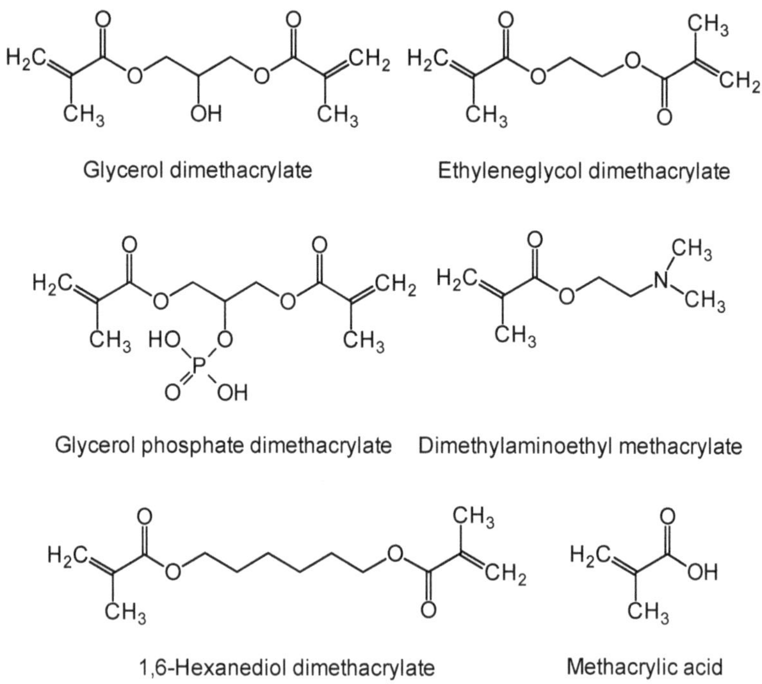

Figure 1.1 (cont.) Monomers used in dental formulations

Table 1.2 (cont.) Acronyms for momomers (4)

Acronym	Chemical Name
HFGA-GMA	Hexafluoroglutaric anhydride-glycerodimethacrylate adduct
HPMA	2-Hydroxypropyl methacrylate
MA	Methacrylic acid
MAEPA	2,4,6-Trimethylphenyl-2-[4-(dihydroxyphosphoryl)-2-oxabutyl]acrylate
MAC-10	11-Methacryloyloxy-1,10-undecanedicarboxylic acid
10-MDP	10-Methacryloyloxydecyl dihydrogen phosphate
MDPB	Methacryloyloxydodecylpyridinium bromide
4-META	4-Methacryloyloxyethyl trimellitate anhydride
4-MET	4-Methacryloyloxyethyl trimellitic acid
MMA	Methyl methacrylate
MMEP	Mono-2-methacryloyloxyethyl phthalate (sometimes also called PAMA Phthalic acid monomethacrylate)
5-NMSA	*N*-Methacryloyl-5-aminosalicylic acid
NPG-GMA	*N*-Phenylglycine glycidyl methacrylate
NTG-GMA	*N*-Tolyl glycine glycidyl methacrylate or *N*-(2-hydroxy-3-((2-methyl-1-oxo-2-propenyl)oxy)propyl)-*N*-tolyl glycine
PEGDMA	Poly(ethylene glycol) dimethacrylate
PEM-F	Pentamethacryloyloxyethylcyclohexaphosphazene monofluoride
PENTA	Dipentaerythritol pentaacrylate monophosphate
Phenyl-P	2-(methacryloyloxyethyl)phenyl hydrogenphosphate
PMDM	Pyromellitic diethyl methacrylate or 2,5-dimethacryloyloxyethyloxycarbonyl-1,4-benzenedicarboxylic acid
PMGDM	Pyromellitic glycerol dimethacrylate or 2,5-bis(1,3-dimethacryloyloxyprop-2-yloxycarbonyl)benzene-1,4-dicarboxylic acid
Pyro-EMA	Tetramethacryloyloxyethyl pyrophosphate
TCB	Butane-1,2,3,4-tetracarboxylic acid di-2-hydroxyethyl methacrylate ester
TEGDMA	Triethylene glycol dimethacrylate
TMPTMA	Trimethylolpropane trimethacrylate
UDMA	Urethane dimethacrylate or 1,6-di(methacryloyloxyethylcarbamoyl)-3,30,5-trimethylhexane

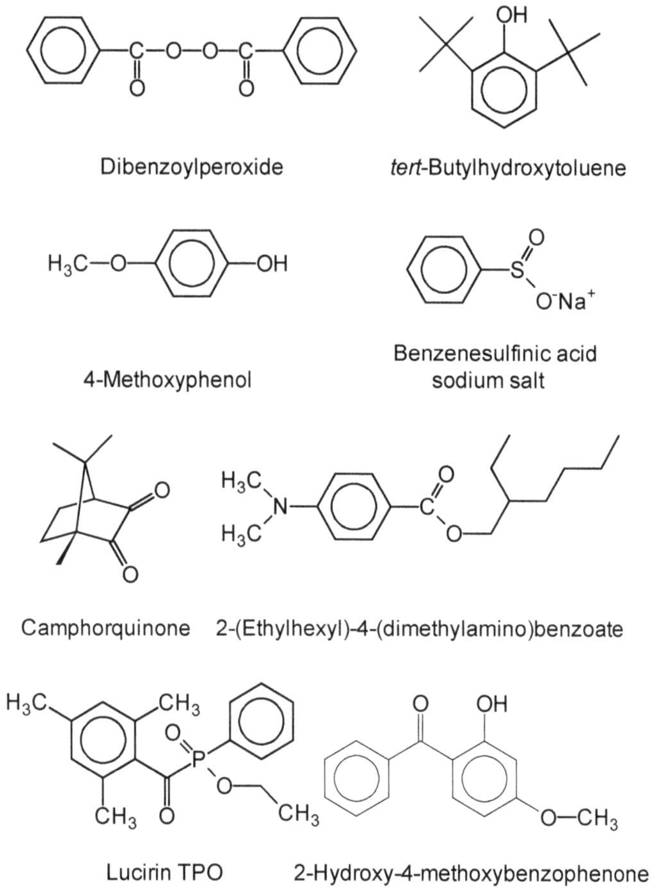

Figure 1.2 Initiators and inhibitors.

Table 1.3 Acronyms for initiators and inhibitors (4).

Acronym	Chemical Name
BHT	Butylhydroxytoluene or butylated hydroxytoluene or 2,6-di-(*tert*-butyl)-4-methylphenol (inhibitor)
BPO	Dibenzoyl peroxide (redox initiator)
BS acid	Benzenesulfinic acid sodium salt (redox initiator)
CQ	Camphorquinone or 1,7,7-trimethylbicyclo-[2,2,1]-hepta-2,3-dione (photoinitiator)
DHEPT	*N,N*-Di-(2-hydroxyethyl)-4-toluidine (co-initiator)
MEHQ	4-Methoxyphenol or monoethyl ether hydroquinone (inhibitor)
ODMAB	2-(Ethylhexyl)-4-(dimethylamino)benzoate (co-initiator)
TPO	Lucirin TPO, BASF (photoinitiator) also ethyl(2,4,6-trimethylbenzoyl)phenylphosphinate
UV-9	2-Hydroxy-4-methoxybenzophenone (photoinitiator)

permanent facings or anterior crowns that may or may not be attached to a metal substructure. It also applies to polymer-based dental crown and bridge materials for which the manufacturer claims adhesion to the metal substructure without macromechanical retention such as beads or wires.

ISO 4049:2009 specifies the requirements for dental polymer-based restorative materials supplied in a form suitable for mechanical mixing, hand-mixing, or intraoral and extraoral external energy activation, and intended for use primarily for the direct or indirect restoration of cavities in the teeth and for luting (9).

The change in color of the facial surface gingiva can be used to determine and optimize the efficacy of antigingivitis treatments (11). Chronic inflammatory disease of the gingiva and periodontium results in destruction of gingival connective tissue, periodontal ligament, and alveolar bone. Clinically, inflammation is seen as redness, swelling, and bleeding observed upon probing. The procedure according to ASTM E2545-07 is suitable for use in diagnosis and monitoring, research and development, epidemiological or other surveys, marketing studies, comparative product analysis, and clinical trials.

ISO 4823:2015 specifies the requirements and tests that the state-of-the-art body of knowledge suggests for helping determ-

Table 1.4 Standards in dentistry.

Standard	Name	Reference
ISO 7491:2000	Dental Materials – Determination of colour stability	(5)
ISO 10477:2004	Dentistry – Polymer-based crown and bridge materials	(6)
ISO 10139-1:2005	Dentistry – Soft lining materials for removable dentures – Part 1: Materials for short-term use	(7)
ISO 9333:2006	Dentistry – Brazing materials	(8)
ISO 4049:2009	Dentistry – Polymer-based restorative materials	(9)
ISO 9693-1:2012	Dentistry – Compatibility testing – Part 1: Metal-ceramic systems	(10)
ASTM E2545-07	Standard Test Method for Objective Measurement of Gingival Color Using Digital Still Cameras	(11)
ISO 4823:2015	Dentistry – Elastomeric impression materials	(12)
ISO 6872:2015	Dentistry – Ceramic materials	(13)
ASTM E3014-15	Standard Practice for Managing Sustainability in Dentistry	(14)
ISO 9693-2:2016	Dentistry – Compatibility testing – Part 2: Ceramic-ceramic systems	(15)
ISO 10139-2:2016	Dentistry – Part 2: Materials for long-term use	(16)
ISO/TS 16506:2017	Dentistry – Polymer-based luting materials containing adhesive components	(17)

ine whether the elastomeric impression materials, as prepared for retail marketing, are of the quality needed for their intended purposes (12).

ISO 6872:2015 specifies the requirements and the corresponding test methods for dental ceramic materials for fixed all-ceramic and metal-ceramic restorations and prostheses (13).

The standard ASTM E3014-15 (14) presents a set of generally recognized activities governing the management of sustainability in dentistry and related dental service practice and a management system framework that assists dentists and dental service organizations to enhance their respective organizational performance and effectiveness.

ISO/TS 16506:2017 specifies test methods and information of bond strength to dentine and physical and chemical performances of dental polymer-based luting materials containing adhesive components (17). The materials are supplied in a form suitable for mechanical mixing or hand-mixing, including using auto-mixing tips, for self-curing and/or external energy activation, or non-mixing for external energy activation.

Additional ISO standards with respect to dental materials can be found online in the ISO standards catalogue (18).

1.2.3 *Adhesion in Restorative Dentistry*

The issues of adhesion in pharmaceutical, biomedical, and dental fields and the theories and mechanisms of adhesion have been described in a monograph (19).

Bonding agents play a crucial role in the effective sealing and retention of resin-based composite restorations, which have been increasingly placed and replaced by dentists (20). Actually, direct adhesive restoration with composite resins has become the procedure of choice for the treatment of anterior and posterior teeth.

However, the long-term durability of those restorations may be compromised due to progressive loss of the integrity of adhesive interfaces. This means that no adhesive strategy is free from technique sensitivity. The specificity and proportion of different constitutive molecules, the interaction between them and substrates can differ greatly from one class of adhesive system to another, which can affect bond quality. An overview of the most important issues in dental adhesion and adhesive systems has been presented, also their composition and clinical use have been discussed (20).

1.2.4 Fracture Toughness

IPS e.max CAD and IPS e.max Press from Ivoclar Vivadent AG are lithium disilicate glass ceramics marketed as interchangeable materials indicated for the same clinical uses (21).

However, different crystal sizes of lithium disilicate are formed during the processing of each of these materials, a factor that could lead to significantly different mechanical properties.

As mechanical failure is always associated with a crack-initiation/crack-propagation process, fracture toughness values could be useful in comparing the different ceramics and possibly predicting clinical performance (21).

The notchless triangular prism specimen K_{ic} test was used to determine and compare the fracture toughness value K_{ic} of the two glass ceramics. A significantly higher K_{ic} value was determined for IPS e.max Press than for IPS e.max CAD. Fractured surfaces, characterized by SEM, showed a marked difference between the two materials, suggesting a more complete crystallization in IPS e.max Press, which was most likely responsible for the higher K_{ic} determined (21).

1.2.5 Biocompatibility of Dental Adhesives

The accomplishment of developing a truly adhesive bond between a restorative material and the natural tooth structures is the goal of adhesive dentistry (22). Dentine adhesive systems come into close contact with dental and oral tissue, especially the pulp and gingival cells.

Due to this close and long-term contact, adhesives should exhibit a high degree of biocompatibility. Biocompatibility is one of the most important properties of dental materials. It has been demonstrated that various components of adhesives can be released (22).

Numerous *in-vitro* investigations have shown that released monomers and other components can cause damage to cultured cells. In addition, many *in-vivo* studies have shown that uncured components which reach the pulpal space cause inflammatory response and tissue disorganization. Only a combination of various *in-vitro* and *in-vivo* tests can provide an overview of the interaction of biomaterials with the host. Therefore, it is necessary on a regular basis to carry out and reverify the biological compatibility of the increasing number of new dental materials. Adhesives should be biofunctional, protective, and preventive, with health-promoting effects that contribute to a better prognosis for restorative treatments and their biocompatibility (22).

1.2.6 *Testing the Cytotoxicity*

The leukocyte viability can be tested using the trypan blue exclusion technique (23). Trypan blue is an azo dye (24). In biosciences, it is used as a vital stain to selectively color dead tissues or cells blue. Trypan blue is shown in Figure 1.3.

Figure 1.3 Trypan blue.

Since cells are very selective in the compounds that can pass through the membrane, trypan blue is not absorbed in a viable cell. However, trypan blue traverses the membrane in a dead cell. Hence, dead cells are shown as a distinctive blue color under a microscope. Since live cells are excluded from staining, this staining method is also described as a dye exclusion method (24).

In the trypan blue exclusion technique, a cell suspension is mixed with 0.4% trypan blue (Sigma) and analyzed with an Olympus light microscope under 100-fold magnification. For each test group, 1000 leukocytes are analyzed by counting the unstained, i.e., viable cells. Blue-colored cells are considered to be nonviable (25,26).

1.2.7 *Degradation of Dental Polymers*

The degradation of dental polymers has been described in a monograph (27). Here also the development of appropriate measuring devices has been reviewed.

The principal modes of dental composite material degradation have been reviewed (28). Particular emphasis has been placed on the selection of the monomer resins, the filler content, and the degree of monomer conversion after the clinical materials are cured.

Also, the loss of the mechanical properties and the leaching of components from the composites have been described. Studies dealing with the chemical breakdown of the materials by agents that are present in the oral cavity have been also reviewed. It could be shown that there are many reasons for the biochemical stability of composite resins in the oral cavity (28).

The potential degradation of monomer systems is dependent on the inherent chemical stability of the atomic groups making up the monomer (28). In the case of dental resins, a resistance to oxidative and hydrolytic mechanisms is important, since there is the potential for both to occur as a result of the exposure of components to salivary fluids.

Basically all dental restorative resin monomers in commercial products are based on the coupling of chemical moieties via ester linkages. However, esters are highly prone to hydrolysis. The predominant resin monomers consist of complexed methacrylate resins. These were introduced as early as 1956 (29). A hybrid molecule was described that was polymerized through methacrylate groups coupled with bisphenol A derivatives via ester groups.

An experimental dimethacrylate monomer was synthesized by the reaction of glycidyl methacrylate (29) and was later produced by the reaction of methacrylic acid and the diglycidyl ether of bisphenol A (30).

The synthesis and the characterization of a series of multifunctional methacrylate-based dental monomers have been described (31). Triethanolamine was reacted with glycidyl methacrylate and methacryloyl chloride to produce a series of multifunctional methacrylate-end-capped compounds for conversion to quaternary ammonium fluoride monomers with decyl-substituted side chain to afford antibacterial dental monomers.

The chemical structure of all samples was characterized by Fourier transform infrared and proton nuclear magnetic resonance spectroscopy. The thus obtained monomers have the potential to replace 2,2-bis[4-(2-hydroxy-3-methacryloyloxypropyl)phenyl]-propane as base monomer of universal resin-based dental composites in the presence of a diluting monomer, e.g., triethylene glycol dimethacrylate, due to their multifunctionality as well as their possible antibacterial activity (31).

1.2.8 *Effect of Modulated Photoactivation on Polymerization Shrinkage*

The influence of modulated photoactivation on axial polymerization shrinkage, shrinkage force, and hardening of light-curing and dual-curing resin-based composites was investigated.

Three light-curing resin composites (SDR bulk-fill, Esthet X flow, and Esthet X HD) and one dual-curing material (Rebilda DC) were subjected to different irradiation protocols with an identical energy density of 27 $J\,cm^{-2}$ (32):

1. High-intensity continuous light (HIC),
2. Low-intensity continuous light (LIC),
3. Soft-start (SS), and
4. Pulse-delay curing (PD).

Axial shrinkage and shrinkage force of specimens with 1.5 mm thickness were recorded in real-time for 15 min. The Knoop hardness was determined at the end of the experimental observation (32).

The Knoop hardness test is a microhardness test used particularly for very brittle materials or thin sheets, where only a small indentation may be made for testing purposes. A pyramidal diamond point is pressed into the polished surface of the test material with a known load for a specified dwell time, and the resulting indentation is measured using a microscope (33).

Statistical analysis revealed no significant differences among the curing protocols for both Knoop hardness and axial shrinkage, irrespective of the composite material (32).

Pulse-delay curing generated the significantly lowest shrinkage forces within the three light-curing materials: SDR bulk-fill, Esthet X flow, and Esthet X HD. High-intensity continuous light created the significantly highest shrinkage forces within Esthet X HD and Rebilda DC, and caused significantly higher forces than LIC within Esthet X flow. Pulse-delay curing decreases the shrinkage forces compared with high-intensity continuous irradiation without affecting hardening and the axial polymerization shrinkage (32).

1.2.9 Ceramics Versus Resin Composites

For posterior partial restorations an overlap of indication exists where either ceramic or resin-based composite materials can be successfully applied. The fatigue resistance of modern dental ceramic materials versus dental resin composites has been compared (34).

Bar specimens of five ceramic materials and resin composites were produced according to ISO 4049 (9) and stored for 14 *d* in distilled water at 37°C. The following ceramic materials were selected for testing: A high strength zirconium dioxide (e.max ZirCAD, Ivoclar), a machinable lithium disilicate (e.max CAD, Ivoclar), a pressable lithium disilicate ceramic (e.max Press, Ivoclar), a fluorapatite-based glass-ceramic (e.max Ceram, Ivoclar), and a machinable color-graded feldspathic porcelain (TriLuxe Forte, VITA). The composite materials selected were: an indirect machinable composite (Lava Ultimate, 3M ESPE) and four direct composites with varying filler nature (Clearfil Majesty Posterior, Kuraray; GrandioSO, VOCO, Tetric EvoCeram, Ivoclar Vivadent; and CeramX Duo, Dentsply).

Fifteen specimens were tested in water for initial strength in 4-point bending. Using the same test setup, the residual flexural fatigue strength was determined using the staircase approach after 104 cycles at 0.5 *Hz* (n = 25) (34).

The zirconium oxide ceramic showed the highest initial strength of 768 *MPa* and flexural fatigue strength of 440 *MPa*. Both lithium disilicate ceramics were similar in the static test. However, the pressable version showed a significantly higher fatigue resistance after cyclic loading. Both the fluorapatite-based and the feldspathic porcelain showed equivalent initial and cyclic fatigue properties. From the composites, the highest filled direct material, Clearfil Majesty Posterior, showed a superior fatigue performance. From all materials, e.max Press and Clearfil Majesty Posterior showed the lowest strength loss of 29.6% and 32%, respectively, whereas the other materials lost between 41% and 62% of their flexural strength after cyclic loading.

In summary, apart from the zirconium oxide and the lithium disilicate ceramics, resin composites generally showed better flexural fatigue strength after 10,000 cycles than the fluorapatite glass-ceramic and the feldspathic porcelain. Thus, resin composite restorations may be used as an equivalent alternative to glass-rich-ceramic inlays with regard to their mechanical performance (34).

1.3 Materials

1.3.1 *Provisional Restoratives*

1.3.1.1 *3D Printed Versus Conventionally Cured Materials*

To optimize the 3D printing of a dental material for provisional crown and bridge restorations a low cost stereolithography 3D printer was used. The mechanical properties were compared with conventionally cured provisional dental materials (35).

For 3D printing a commercial printable resin (NextDent C&B Vertex Dental) was used. Two conventionally cured provisional materials (Integrity®, Dentsply and Jet®, Lang Dental Inc.) were used.

NextDent C&B is a biocompatible printable material for bridges and crowns (36).

The results of this study suggest that a 3D printable provisional restorative material allows for sufficient mechanical properties for intraoral use, despite the limited 3D printing accuracy of the printing system of choice (35).

1.3.1.2 *Color Stability*

The color stability of provisional restorative materials after storing in different drinks was compared (37). Thirty specimens with 10 *mm* diameter and 2 *mm* thickness were prepared from three different materials (Temdent, TRIAD ve Telio CAD).

Telio CAD are crosslinked PMMA blocks for the efficient fabrication of long-term temporaries (38).

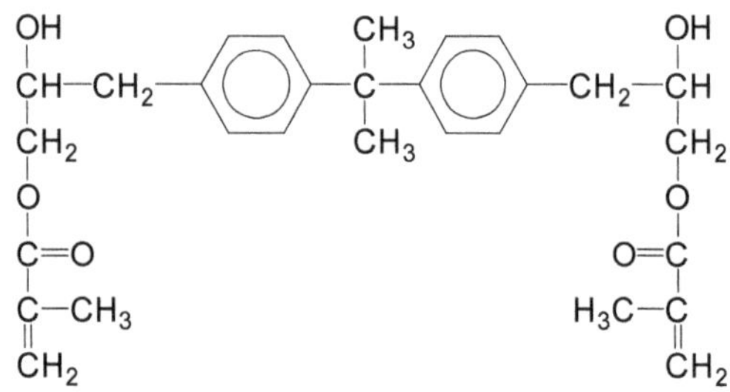

2,2-Bis[*p*-(2-hydroxy-3-methacryloxypropoxy) phenyl]propane

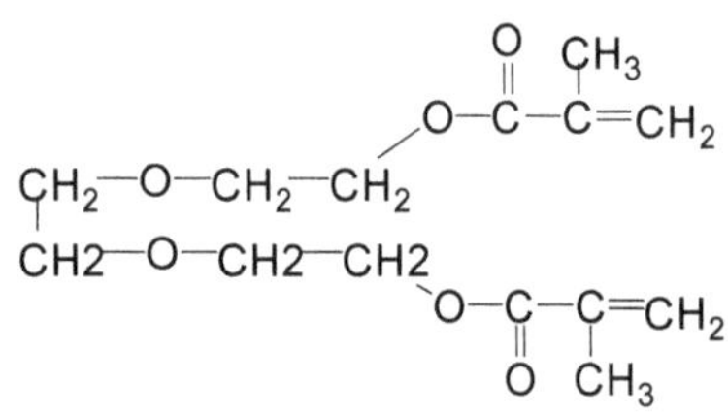

Triethylene glycol dimethacrylate

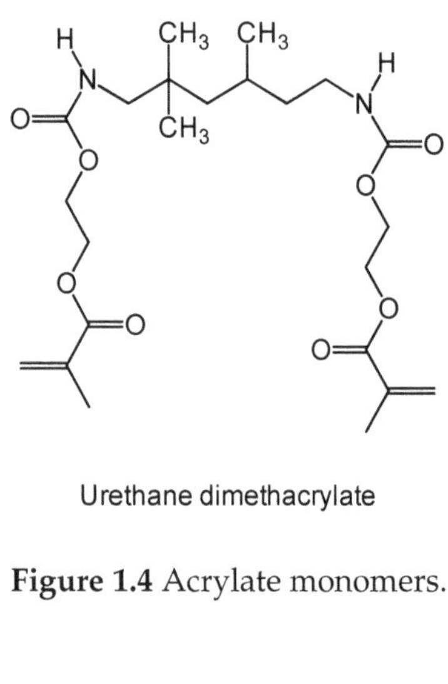

Urethane dimethacrylate

Figure 1.4 Acrylate monomers.

Table 1.6 Photopolymerization initiators (39).

Compound
Benzoin alkyl ethers
Benzoin methyl ether Benzoin ethyl ether Benzoin isopropyl ether
Benzyl ketals
Benzyl dimethyl ketal Benzyl diethyl ketal
Diaryl ketones
Benzophenone Anthraquinone Thioxanthone
α-Diketones
Diacetyl Benzil Camphorquinone 9,10-Phenanthraquinone
Bisacylphosphine oxides
Bis-(2,6-dichlorobenzoyl)phenylphosphine oxide Bis-(2,6-dichlorobenzoyl)-2,5-dimethylphenylphosphine oxide Bis-(2,6-dichlorobenzoyl)-4-propylphenylphosphine oxide Bis-(2,6-dichlorobenzoyl)-1-naphthylphosphine oxide Bis(2,4,6-trimethylbenzoyl)-phenylphosphine oxide

Table 1.7 Reducing compounds (39).

Compound
2-(Dimethylamino)ethyl methacrylate Ethyl 4-(dimethylamino)benzoate *N*-Methyldiethanolamine Dimethylaminobenzaldehyde Terephthalaldehyde 2-Mercaptobenzoxazole 1-Decanethiol Thiosalicylic acid Thiobenzoic acid *N*-Phenylalanine

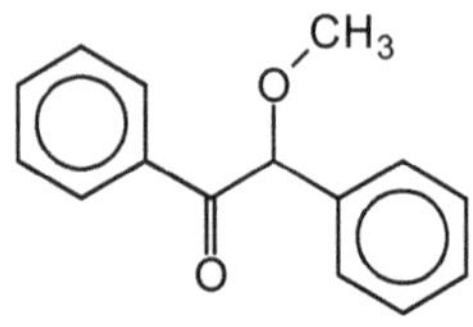

Benzoin methyl ether

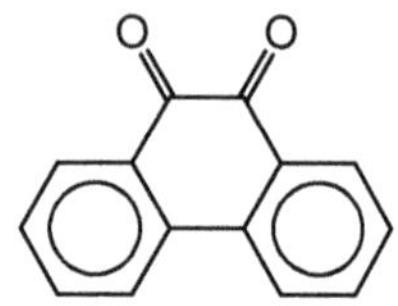

9,10-Phenanthraquinone

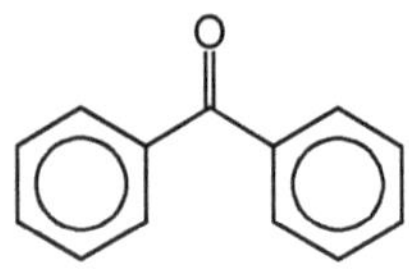

Benzophenone

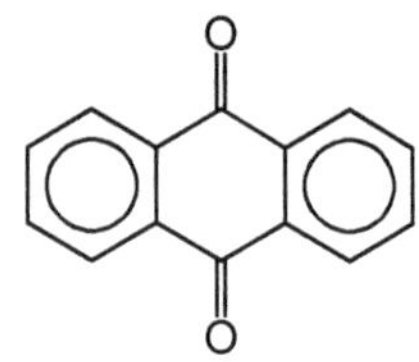

Anthraquinone

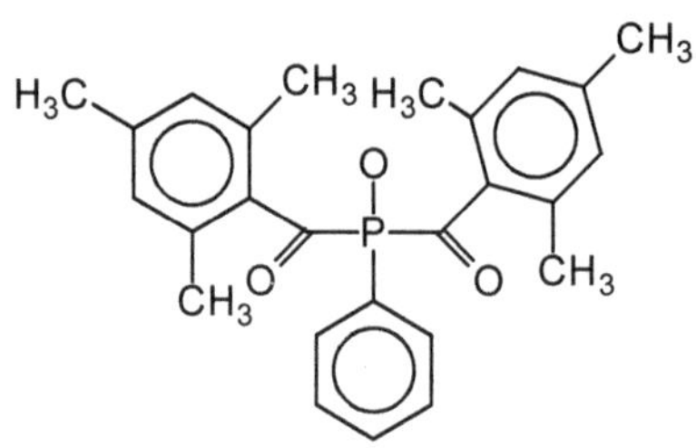

Bis(2,4,6-trimethylbenzoyl)-phenylphosphine oxide

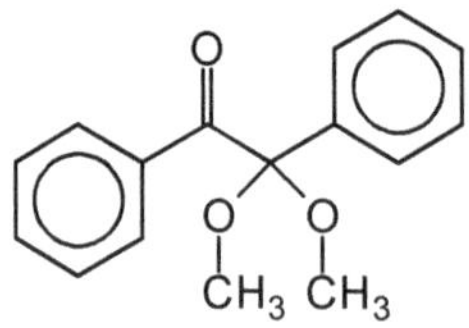

Benzyl dimethyl ketal

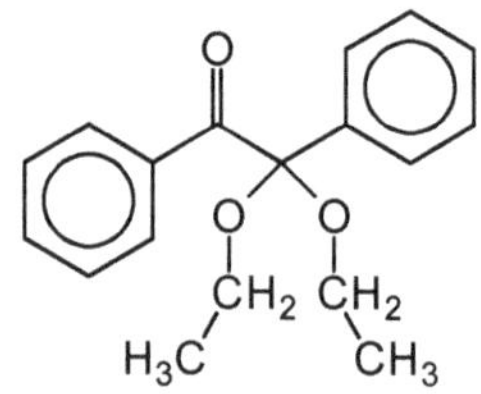

Benzyl diethyl ketal

Figure 1.5 Photopolymerization initiators.

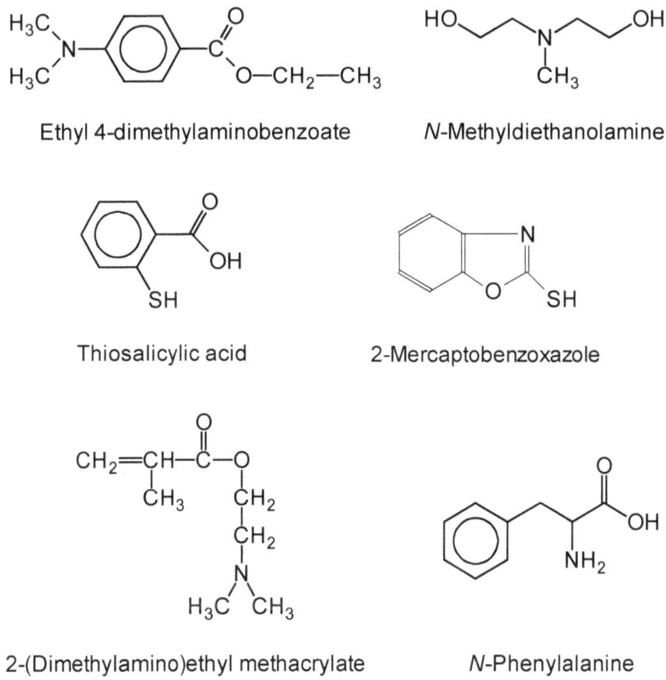

Figure 1.6 Reducing compounds (39).

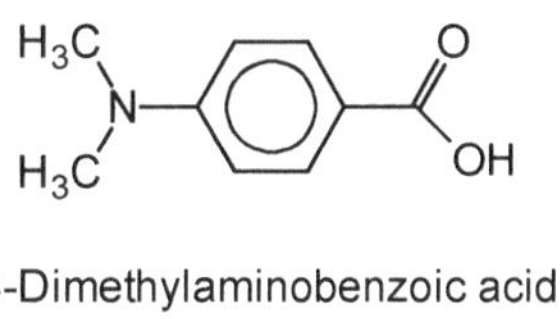

Figure 1.7 Reducing compound (39).

aluminium fluoro-silicate glass fillers can optionally be treated with an additional organic compound.

Additional organic compounds contain ethylenically unsaturated groups and hydrophilic groups, such as, for example, ethylene glycol groups.

A process for the preparation of a filler based on glass particles has been described. Here, glass particles are dispersed in a solution of a homo- or copolymer of vinyl chloride in an organic solvent and the solvent is then removed (42).

It was found that the PVC on the surface of the glass particles of the filler when the latter is used in an acid organic matrix, i.e., in a self-adhesive composite, seriously impedes the dissolving of cations, e.g., Ba^{2+} ions, by the acid constituents of the matrix; and thus a salt formation and viscosity increase of the organic matrix is reduced and finally a significant improvement of the storage stability of the pasty dental materials can be achieved (42).

This is surprising, in as much as the release, for example, of fluoride ions is not prevented, or is actually improved, by the surface treatment of fillers with polymeric organic compounds (42).

1.3.6 Glass Ionomers

Glass-ionomer cements for use in clinical dentistry have been reviewed (45). Glass-ionomer cements belong to the class of materials known as acid-base cements. They are based on the product of reaction of weak polymeric acids with powdered glasses of basic character. The setting reaction occurs in concentrated solutions in water and the final structure contains a substantial amount of unreacted glass which acts as filler to reinforce the set cement.

The ISO requirements of glass ionomer cements are summarized in Table 1.8.

Glass ionomers used as dental restoratives have several advantages over conventional resin-containing composites (47). For example, glass ionomers are tolerant to their application on wet surfaces, have low shrinkage and are self-adhesive. Since glass ionomers contain polymers rather than monomers, there is no risk of acrylic monomers leaching out, which can lead to sensitization and allergic reactions. Furthermore, glass ionomers bond chemically to dental

Table 1.8 ISO requirements of glass ionomer cements (45,46).

Property	Luting Cement	Restorative Cement
Setting time /[*min*]	2.5–8	2–6
Minimum compressive strength /[*MPa*]	70	100
Maximum acid erosion /[$mm\,h^{-1}$]	—	0.05
Opacity, $C_{0.70}$	−0.35	−0.90
Acid-soluble arsenium (As) /[$mg\,kg^{-1}$]	2	2
Acid-soluble lead (Pb) /[$mg\,kg^{-1}$]	100	100

hard tissues, and may also provide a beneficial level of fluoride release, which helps to prevent recurrent caries. Accordingly, ionomer cements are widely used in the dental field for filling of a cavity, cementing of crowns, inlays, bridges, or orthodontic bands, lining of a cavity, sealing of a root canal, core construction, and preventive sealing.

In glass ionomer cements, the primary reactions which cause the glass ionomer cement to harden is crosslinking based on ionic forces between metal ions released from the glass and the polymer comprising acidic groups. Moreover, the acids of the glass ionomer cement partially dilute metal cations from the glass structure during setting so that ionic carboxylates of metal cations may be formed during the setting process (47).

A key weakness of commercial glass ionomers, however, is their low flexural strength manifesting itself as an undesirable brittleness, which may lead to fracture at the edges of a restoration and, in the worst case, to bulk fracture of a restoration (47). Therefore, the restorative application of ionomer cements in posterior teeth is usually limited to non-stress-bearing areas. Ionomer cement materials continue to have significant limitations for use in permanent posterior restorations, particularly with regard to large restorations.

These problems have been solved with an aqueous dental glass ionomer composition containing (47):

1. A reactive particulate glass,
2. A linear or branched polymer comprising acidic groups, which is reactive with the particulate glass in a cement reaction, whereby the linear or branched polymer comprising

from poly(acrylic acid) or copolymers of acrylic and itaconic acid, acrylic acid and maleic acid (47), cf., Figure 1.12.

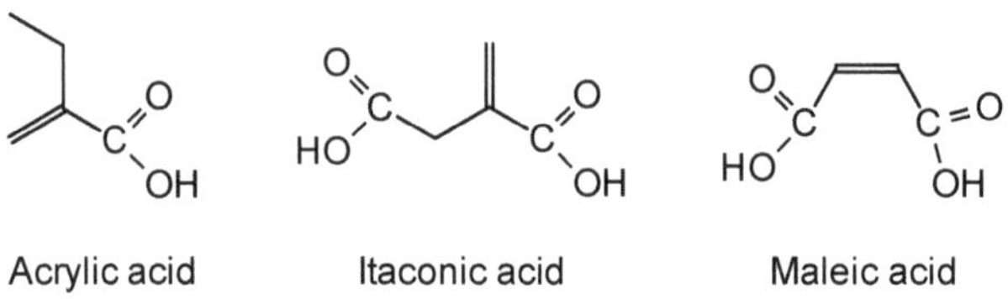

Figure 1.12 Comonomers with acidic groups.

1.3.7 *Bioactive Glass*

Bioactive glass is known to possess antimicrobial properties and can release ions needed for remineralization of tooth tissue (48).

Bioactive glasses are able to bond to both soft and hard tissue and promote the bone growth. The bioactivity behavior of these glasses is related to the formation of a biologically active hydroxyapatite layer on the surface of the glasses. The mechanism of bonding of bioactive glasses to tissues includes a series of surface reactions that occur when the glass is exposed to an aqueous environment (49).

One of the most important properties of bioactive glasses is their ability to exhibit antibacterial activity, which creates a bacteria-free environment while healing and regenerating the defect area. These potentials of bioactive glass make it a unique material to be widely used in dentistry. Such materials can stimulate bioactive behavior around the fixed restorations margins and provide a bioactive surface. Therefore, they can develop periodontal tissue attachment and create complete sealing of the marginal gap. This sealing can prevent the failure of fixed ceramic restorations by eliminating the secondary caries, micropenetration of oral bacteria and their adhesion on the cement surface (49).

Bioactive glass has been developed for dental restorative composites with adequate mechanical properties. Composites with a content of 72% total filler were prepared while substituting 0–15% of the filler with a ground bioactive glass (48).

The flexural strength, fracture toughness, and fatigue crack growth resistance for the composites containing bioactive glass were found to be unaffected by increasing the bioactive glass content up to 15% and were superior to Heliomolar after all post-cure treatments. Heliomolar is a product from Ivoclar Vivadent Corp.

Furthermore, the flexural strength of the bioactive glass composites was unaffected by two months exposure to aqueous media and a bacterial challenge, but some decreases in the fracture toughness and the fatigue resistance were observed.

The favorable mechanical properties compared to Heliomolar were attributed to the higher filler content and a microstructure morphology that better promoted the toughening mechanisms of crack deflection and bridging (48).

Also, the physical and antibacterial properties of a flowable resin composite were studied with a sol-gel derived silver-doped bioactive glass (50). The depth of the cure was calculated by measuring the surface microhardness for the top and bottom surfaces. The volumetric polymerization shrinkage was measured by recording the linear shrinkage as change in length, while the biaxial flexural strength was studied measuring the load at failure. The antibacterial properties of the samples were tested against *Streptococcus mutans* and *Lactobacillus casei*.

The thus obtained values were slightly decreased for all tested physical properties in comparison to those of a control group, a flowable resin composite without Ag bioactive glass

However, an enhanced bacteria inhibition was observed for the composite. This composite could find an application in low stress-bearing areas as well as in small cavity preparations to decrease secondary caries (50).

The influence of biofilm formation on the optical properties of bioactive glass-containing composites were elucidated (51). A bioactive glass and fluoride-containing bioactive glass composites were prepared. Then, the effects of biofilm on the optical properties of these composites were investigated. The biofilm may contribute to a decrease of the color change of these composites, since all the color differences for the composites showed significant decreases with a bacterial biofilm (51).

Flame spray synthesis has opened the possibility to add additional elements to complex materials, such as bioactive glasses, while maintaining nanoparticulate properties (52). It was investigated whether a flame-sprayed bismuth oxide doped nanome-

A light-curable composition is shown in Table 1.11. Diurethane dimethacrylate is shown in Figure 1.14.

Table 1.11 Light-curable composition (57).

Ingredient	[%] per weight
Diurethane dimethacrylate	73.75
Poly(butene)	8.38
Camphorquinone	0.75
Methacryloyl ethyl diethylamine	1.06
Silica	16.86

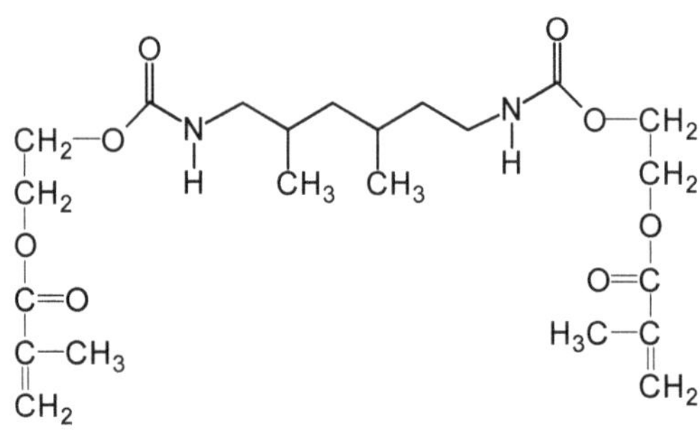

Figure 1.14 Diurethane dimethacrylate.

The materials in Table 1.11 were mixed and then cured for 20 s using an Optilux dental curing device (57).

A chemically curable composition consists of a 1:1 per volume mixture of Parts A and B, as shown in Table 1.12.

The material was soft cured within 110 s and hard cured in 150 s at 23°C. Soft curing is a stage at which the material becomes solid but still exhibits flexibility. The properties of the cured material are: Barcol hardness 42–45 and flexural strength 61 MPa (57).

1.3.10 *Hydrophilic Polymer Sponge Structures*

After a tooth extraction, the quick cessation of bleeding and the formation of a blood clot on the wound in the open tooth socket are very desirable (58). During the entire healing period following an extraction it is important to preserve conditions conducive to

Table 1.12 Chemically curable composition (57).

Ingredient	[%] per weight
Part A	
Diurethane dimethacrylate	78.96
N,*N*-Bis(2-hydroxyethyl)-*p*-toluidine	0.39
Silica	20.65
Part B	
Diurethane dimethacrylate	73.58
Poly(butene)	9.68
BHT	0.05
Dibenzoyl peroxide	1.06
Silica	15.63

hemostasis, so that the blood clot that forms within the socket does not break down and/or dislodge. If the clot breaks down and/or dislodges, a condition known as dry socket, also called alveolar osteitis, results. Dry socket conditions can also occur for the same reason during the treatment of cystic cavity defects in the jaw. Dry socket can cause pain and discomfort, which will subside only as the socket heals through a secondary healing process.

Systems and methods for treating tissue or bone in an oral cavity or an adjacent anatomic structure have been described which consist of the placement of a hydrophilic polymer sponge structure. Assemblies for treating tissue or bone in an oral cavity or an adjacent anatomic structure have been described (58, 59). The systems and methods utilize hydrophilic polymer sponge structures that will be placed within the area that will receive the gingival graft material.

Dental dressing assemblies have been described that are formed from hydrophilic polymer sponge structures, such as a densified chitosan biomaterial (59). The enhanced physical properties of the densified chitosan matrix are enhanced by the adhesive strength of the chitosan matrix, its self-promotion of coagulation, and its antibacterial, antimicrobial, and anti-viral properties.

1.3.11 *Hybrid Polymer Network Compositions for Artificial Teeth*

Recently, plastic teeth have largely eliminated porcelain teeth from the denture tooth market due to various advantages, such as better bond to the denture base, lighter weight, toughness, less undesirable noises during chewing, and less wear to the opposing natural or artificial teeth, crown, or bridge. However, plastic teeth have the disadvantage of being more subject to wear than porcelain teeth (60).

Among the presently available organic compositions used for the construction of artificial teeth, most are composed from acrylic materials. While such compositions are commonly used for artificial teeth, they possess certain drawbacks. In general, artificial teeth made of currently available acrylic compositions do not have sufficient wear resistance and can be deformed by relatively low biting forces. The deficiency in wear resistance and limited loading capability of current polymeric artificial teeth are apparent when they are compared to natural or ceramic teeth, crowns or bridges.

Polymeric compositions comprising crosslinked polymers, and highly crosslinked polymers, monomers, and multifunctional crosslinking monomers or oligomers for these monomers, are capable of being formed or molded and polymerized to provide articles possessing superior wear resistance, physical and physicochemical properties. In addition, these compositions have excellent molding properties and processing capabilities which are useful in the production of artificial teeth, dental crowns, or other dental restorations (60).

A highly crosslinked polymer powder has been prepared from a composition shown in Table 1.13.

Table 1.13 Composition for highly crosslinked polymer powder (60).

Ingredient	[%] per weight
Methyl methacrylate	25.0
2,2-Bis(4-methacryloxyphenyl)propane	15.5
1,6-Hexanediol dimethacrylate	14.0
1,4-Cyclohexanediol dimethacrylate	45.0
Dibenzoyl peroxide	0.5

Also, several other slightly modified compositions have been described (60).

1.3.12 *Urethane Macromers*

A urethane macromer containing hexafluoroisopropylidene, poly(ethylene oxide) and carboxylic moieties (UF-DMA) was synthesized and used in proportions varying between 15% and 35% in dental adhesive formulations besides Bis-GMA, triethylene glycol dimethacrylate and 2-hydroxyethyl methacrylate (61).

The morphology of the crosslinked networks was compared with specimens containing 10% hydroxyapatite or calcium phosphate. The degrees of conversion after 180 *s* of irradiation with visible light ranged from 59.5% to 74.8%. The water sorption was found to be between 23.15 $\mu g\, mm^{-3}$ and 40.52 $\mu g\, mm^{-3}$. Due to the addition of hydroxyapatite or calcium phosphate this parameter attained values of 37.82–49.14 $\mu g\, mm^{-3}$ and 34.58–45.56 $\mu g\, mm^{-3}$, respectively.

The above-mentioned findings suggest that UF-DMA taken as a comonomer in dental adhesives of the acrylic type may provide improved properties in the moist environment of the mouth (61).

1.3.13 *Catechol-Functionalized Polymer*

A synthetic polymer functionalized with catechol groups as dental adhesives has been described (62). The monomer is shown in Figure 1.15.

This synthetic polymer mimics mussel adhesive proteins that enable mussels to anchor to a variety of wet surfaces. The catechol groups of adhesion proteins displace tightly bound water.

It is believed that a catechol-functionalized polymer functions as a dental adhesive for wet dentin surfaces, potentially eliminating the complications associated with saliva contamination (62).

The mechanical and adhesive properties of the polymer to dentin surface in the presence of water and salivary components were elucidated. It was found that this polymer combined with an Fe^{3+} additive improved the bond strength of a commercial dental adhesive (BeautiBond™ and Scotchbond™ Multi-Purpose) to artificial saliva contaminated dentin surface as compared to a control sample

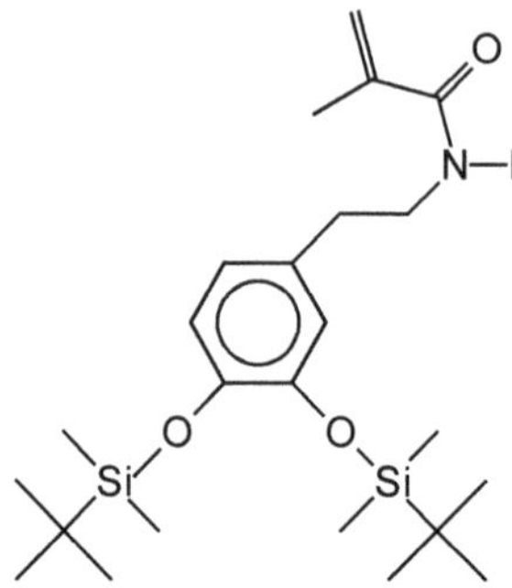

tert-Butyldimethylsiloxy dopamine methacyrlate

Figure 1.15 *tert*-Butyldimethylsiloxy dopamine methacrylate (62).

without the polymer. Histological analysis of the bonding structures showed no leakage pattern, probably due to the formation of Fe-catechol complexes, which reinforce the bonding structures. A cytotoxicity test showed that the polymers did not inhibit the proliferation of human gingival fibroblast cells. The results from this study suggest a potential to reduce failure of dental restorations due to saliva contamination using catechol-functionalized polymers as dental adhesives (62).

1.3.14 High Refractive Index Monomers

Translucency of a dental material is useful for esthetic reasons and for efficient UV curing of the resin (63). To achieve a good translucency, it is desirable to minimize the scattering of light as it passes through the material. This may be accomplished by matching the average refractive index of the filler and the resin.

It has been observed that the incorporation of high refractive index groups increases the depth of cure during a UV/Vis-initiated polymerization process by reducing light scattering due to mismatched refractive indices.

High refractive index monomers have been investigated that may be blended with dental resins to raise the refractive index.

These monomers are isocyanurate-based compounds attached to polymerizable groups (63).

Useful high refractive functional groups are shown in Table 1.14 and in Figure 1.16.

Table 1.14 High refractive functional groups (63).

Group
Phenyl
Benzyl
Biphenyl
Fluorenyl
4-(1-Methyl-1-phenethyl)phenoxyethyl
Phenylthio
Napthyl
Naphthylthio
2,4,6-Tribromophenoxy
2,4-Dibromophenoxy
2-Bromophenoxy
Naphthyloxy
3-Phenoxy
4-Phenylphenoxy
2,4-Dibromo-6-*sec*-butylphenyl
2,4-Dibromo-6-isopropylphenyl
2,4-Dibromophenyl
Pentabromobenzyl Pentabromophenyl

1.3.15 *Antibacterial Polymers for Dental Adhesives*

Antibacterial polymers for dental adhesives and composites have been described in a monograph (64).

Tooth caries are the result of a dietary carbohydrate-modified bacterial infectious disease, one of the most common bacterial infections in humans (65,66).

The basic mechanism of dental caries is demineralization, or mineral loss, through attack by acid generated by bacteria (67,68).

Therefore, the growth of acidogenic bacteria and the biofilm formation are responsible for dental caries (69).

The formation of plaque has been described to have three steps (66):

1. Pellicle formation,

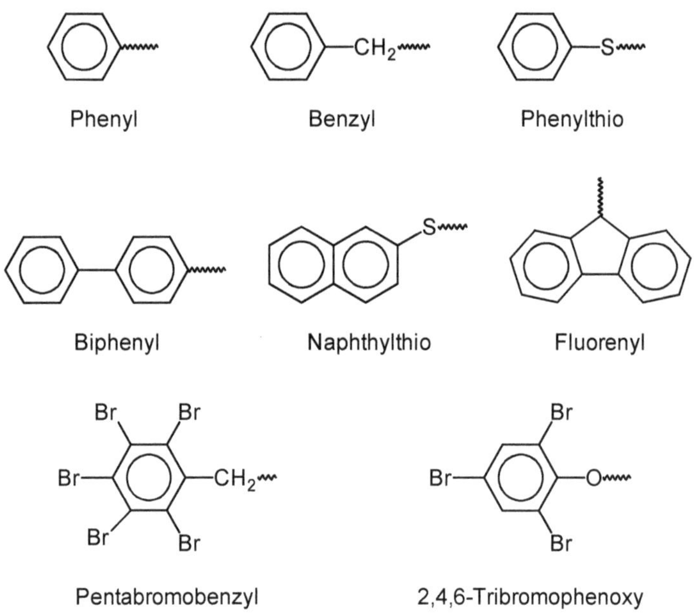

Figure 1.16 High refractive functional groups.

2. Bacteria colonization, and
3. Biofilm maturation.

In the initial stage, a proteinaceous film called pellicle forms on the tooth surface with adsorbed components from saliva, mucosa, and bacteria. Bacteria then adhere and colonize on this surface to grow into a biofilm, which is a heterogeneous structure consisting of clusters of various types of bacteria embedded in an extracellular matrix (70).

Cariogenic bacteria, such as *Streptococcus mutans* and lactobacilli in the plaque, can take nutrients from carbohydrates and produce organic acids. Acid production causes a demineralization of the tooth structure beneath the biofilm.

Antibacterial polymers for dental adhesives and composites have been described in a monograph (71) and reviews have also been presented (72,73).

Also, multifunctional compositions have been investigated (74). It could be demonstrated that dental restorative materials containing multiple therapeutic agents of different chemical characteristics can be beneficial toward improving the resistance to mechanical and acidic challenges in oral environments.

An antibacterial material should be able to stop the formation of a biofilm for a long period by killing infectious bacteria selectively without affecting the normal flora.

The antibacterial monomer 12-methacryloyloxydodecylpyridinium bromide and an adhesion-promoting phosphoric monomer were incorporated into Bis-GMA-based dental resin and its antibacterial activity after curing was investigated (75). The resin that contained 12-methacryloyloxydodecylpyridinium bromide and 10-methacryloyloxydecyl dihydrogen phosphate was polymerized and washed with methanol. Then, the bacteriostatic and bactericidal effects against *Streptococcus mutans* were determined. The growth of *S. mutans* was strongly inhibited by the contact with the surface of the cured 12-methacryloyloxydodecylpyridinium bromide/10-methacryloyloxydecyl dihydrogen phosphate-containing resin, although the bactericidal effect was small.

A cured 12-methacryloyloxydodecylpyridinium bromide/10-methacryloyloxydecyl dihydrogen phosphate-containing resin also showed an inhibitory effect against *in-vitro* plaque formation on its surface by *S. mutans*. The bactericide immobilized in Bis-GMA-based resin demonstrated a bacteriostatic activity as a contact antimicrobial agent even when the adhesion-promoting phosphoric monomer was incorporated into the materials (75).

A quaternary ammonium monomer, dimethylamino dodecyl methacrylate and nanoparticles of silver were incorporated into a primer and an adhesive (76). Their effects on the antibacterial and dentin bonding properties were investigated.

It twas found that the bonding agents are promising to combat residual bacteria in the tooth cavity and invading bacteria at tooth-restoration margins to inhibit caries. Thus, dimethylamino dodecyl methacrylate and nanoparticles of silver are promising materials for their use in a wide range of dental adhesive systems and restoratives (76).

A method has been presented for the preparation of a blend of antibacterial quaternary ammonium silanes and quaternary ammonium methacryloxysilane based on the sol-gel reaction between dimethyldiethoxysilane and two trialkoxysilanes, one with an antibacterial quaternary ammonium functionality and the other with a methacryloxy functionality (77).

The antibacterial adhesive composition killed bacteria in *S. mutans* biofilms not only through the release of the non-copolymerizable quaternary ammonium silane species, i.e., release-killing, but also via immobilized quaternary ammonium methacryloxysilane that are copolymerized with adhesive resin comonomers, i.e., contact-killing (77).

Antibacterial monomers have been described that differ in their alkyl chain length (66). These monomers are shown in Table 1.15.

The synthesis of the quaternary ammonium salt monomethacrylates was done using the Menschutkin reaction (78,79). The reaction proceeds by the addition reaction of tertiary amines with organohalides. To form a quaternary ammonium salt with a reactive methacrylate group, 2-(dimethylamino)ethyl methacrylate was chosen as methacrylate-containing tertiary amine (66).

Because dimethylamino dodecyl methacrylate exhibited a much greater antibacterial potency than dimethylamino hexyl methacrylate and bis(2-methacryloyloxy-ethyl) dimethyl ammonium bromide, dimethylamino dodecyl methacrylate was used for incorporation into a nanocomposite from nanoparticles of amorphous calcium

Table 1.15 Antibacterial monomers (66).

Monomer	Acronym
Dimethylamino propyl methacrylate	DMAPM
Dimethylamino hexyl methacrylate	DMAHM
Dimethylamino heptyl methacrylate	DMAHPM
Dimethylamino octyl methacrylate	DMAOM
Dimethylamino nonyl methacrylate	DMANM
Dimethylamino decyl methacrylate	DMADM
Dimethylamino undecyl methacrylate	DMAUDM
Dimethylamino dodecyl methacrylate	DMADDM
Dimethylamino tridecyl methacrylate	DMATDM
Dimethylamino tetradecyl methacrylate	DMATTDM
Dimethylamino pentadecyl methacrylate	DMAPDM
Dimethylamino hexadecyl methacrylate	DMAHDM
Dimethylamino heptadecyl methacrylate	DMAHPDM
Dimethylamino octadecyl methacrylate	DMAODM
Dimethylamino nonadecyl methacrylate	DMANDM
Dimethylamino icosyl methacrylate	DMAIOM
Dimethylamino henicosyl methacrylate	DMAHOM
Dimethylamino docosyl methacrylate	DMADOM

phosphate. As basic monomers, bisphenol glycidyl dimethacrylate and triethylene glycol dimethacrylate were used (66).

Increasing the mass fraction of dimethylamino dodecyl methacrylate in the nanocomposite resulted in much more red/yellow/orange staining, indicating that these nanocomposites effectively inhibited the biofilm growth.

These results also indicate that the nanoparticles of amorphous calcium phosphate are not antibacterial by themselves, and dimethylamino dodecyl methacrylate is responsible for the antibacterial activity (66).

1.3.16 *Chemical Composition of Contemporary Dental Adhesives*

Dental adhesives are designed to bond composite resins to enamel and dentin. Their chemical formulation determines to a large extent their adhesive performance (4). An adhesive system typically contains resin monomers, curing initiators, inhibitors or stabilizers, solvents and sometimes inorganic filler. Each one of these components has a specific function. The adhesives may be classified into two main groups, etch and rinse adhesives and self-etch adhesives.

Also, chemical formulations of dental adhesives commercially available up to 2007 have been collected (4).

The dentine bond strength of fluoride-containing dental adhesives has been reviewed (80).

Methacrylamide monomers have been proposed as a hydrolytically stable alternative for the formulation of dental adhesives (81). The kinetics of polymerization and microtensile bond strength stability of the adhesives have been assessed.

The fundamental processes responsible for the aging mechanisms involved in the degradation of resin-bonded interfaces and the potential approaches to prevent and counteract this degradation have been reviewed (4).

The resin-dentin bond degradation is a complex process, involving the hydrolysis of both the resin and the collagen fibril phases contained within the hybrid layer. Collagen fibers become vulnerable to mechanical and hydraulic fatigue, as well as degradation by host-derived proteases with collagenolytic activity, i.e., matrix metalloproteinases and cysteine cathepsins. The inhibition of the collagenolytic activity and the use of crosslinking agents are the two main strategies to increase the resistance of the hybrid layer to enzymatic degradation (4).

1.3.17 *Nanostructured Antibacterial and Remineralizing Dental Bonding Agents*

Dental bonding systems that possess antibacterial properties can be useful in inhibiting the development of secondary caries by preventing the growth of residual and invading bacteria. Such antibacterial dental bonding systems can also find use in other applications in which dental bonding is employed, including (82):

- Dentin bonding,
- Enamel bonding,
- Tooth roots,
- Marginal repair,
- As a crown cement,
- As an inlay cement,

- As an onlay cement,
- As a pit and fissure sealant, and
- As an orthodontic bracket adhesive or cement.

The dental bonding agents include dental primers and dental adhesives. The dental primers and the dental adhesives should have antibacterial properties. These properties are imparted by antibacterial agents. In addition, the dental primers and adhesives may have remineralizing properties (82).

The Scotchbond™ Multi-purpose adhesive was used as the parent adhesive to test the effect of the incorporation of bis(2-methacryloyloxy-ethyl) dimethyl ammonium bromide and -ilver containing nanoparticles (82).

The Scotchbond Multi-purpose adhesive etchant contained 37% phosphoric acid, in addition to 35%–45% 2-hydroxyethyl methacrylate, 10%–20% copolymer of acrylic/itaconic acids, and 40%–50% water. The Scotchbond Multi-purpose adhesive contained 60%–70% bisphenol A diglycidyl methacrylate and 30%–40% 2-hydroxyethyl methacrylate. Furthermore, quaternary ammonium dimethacrylate and nanoparticles of silver were incorporated into the adhesive primer. As remineralizing agent, nanoparticles of amorphous calcium phosphate were added.

Several specimens were fabricated and tested as reported in detail (82). Specimens with silver-containing nanoparticles and amorphous calcium phosphate reduced the colony-forming units by an order of magnitude, compared to a control specimen. Specimens with 40% amorphous calcium phosphate slightly reduced the colony-forming units, compared to a sample without amorphous calcium phosphate; however, this decrease was not statistically significant (82).

1.3.18 *Rechargeable Calcium Phosphate-Containing Dental Materials*

Calcium phosphate dental resins and composites used in restorations can release calcium and phosphate ions to inhibit caries, remineralize tooth lesions, and regenerate the lost minerals in tooth structures (83). However, the release of these ions lasts for only a few months, after which the release diminishes and gradually stops.

- Dentures,
- Dental implants composed of a fixture and an abutment, and
- Fixed braces, i.e., train tracks.

These devices for dental restoration must be custom-shaped for an individual and are made by casting. For this reason, the castability of alloys is an important requirement for dental applications (85).

Dental alloys can be grouped into precious and nonprecious metals (85). The particular alloys are selected according to the intended use. Alloys of precious metals, such as gold, palladium, and silver, are usually employed because of their high corrosion resistance, biocompatibility, and castability, in comparison to those of nonprecious metals. Precious alloys are grouped into high-carat alloys and low-carat alloy. A high-carat alloy should contain more than 75 % precious metals. Nonprecious metal alloys are stainless steels, cobalt-chromium, nickel-chromium, and titanium alloys (85).

1.3.20 *Tooth Desensitizing Oral Care Compositions*

Dentinal hypersensitivity is a condition where dentin is exposed due to the lack of soft tissues covering the surfaces (86). The exposed dentin includes small tubules which are susceptible to triggering of a pain response from a variety of stimuli such as heat, cold, sour taste, or pressure. The level of pain can range from an ache or soreness to a shooting pain.

The most common approach to the treatment of dentinal hypersensitivity is the use of dentifrices that contain desensitizing agents such as potassium nitrate and bioactive glass. Such products can provide some relief; however, the reduction of the sensitivity through use of a dentifrice alone is limited for several reasons, including: Inherent limitations on the contact time of the dentifrice to the exposed dentin (as brushing time may be relatively short) and dilution of the desensitizing agent by saliva.

Desensitizing oral care compositions are useful for treating dentinal hypersensitivity. Oxalate salts can act as dentinal tubule blockers that can therefore be useful for the treatment of dental hypersensitivity. Potassium oxalate is a preferred desensitizing agent as it forms calcium oxalate upon delivery to the open tubules. The formation of the calcium oxalate blocks the dentinal tubules and may prevent environmental stimuli from causing sensitivity pain.

An example for such a composition is shown in Table 1.17.

Table 1.17 Tooth desensitizing oral care composition (86).

Ingredient	[%] per weight
Carboxymethyl cellulose 7M8SF	5.0
Glycerin USP (99.7)	31.86
Carbopol 956	1.0
Sodium Benzoate, NF FCC	0.50
Potassium Sorbate	0.20
Sodium hydroxide solution 50	0.25
Potassium oxalate monohydrate	3.14
Purified Water USP	58.05

1.3.21 *Commercially Available Compositions*

The two-body wear resistance and the hardness of three modern composite materials were investigated (87). The composites were Filtek Silorane, Ivoclar Heliomolar and Voco Grandio. Five pieces were produced from each material as recommended by the manufacturers.

All the samples were kept in distilled water for 1 week and the Vickers hardness was determined before the two-body wear test. Then, two-body wear tests were performed using a computer controlled chewing simulator.

The chewing simulator was programmed to provide a 2 *mm* vertical movement and a 0.7 *mm* horizontal movement. For each wear test, Al_2O_3 with a diameter of 6 *mm* was used as antagonist material. The mean volume loss of all samples after the wear tests was determined using three-dimensional profilometry. In addition. a random specimen was selected from each test group and scanning electron microscope images were taken for analysis of wear tracks.

The hardness values of the materials tested were between 49 and 78 Vickers hardness. The lowest mean volume loss was detected for the Heliomolar sample at about 3.1 μm^3 and the highest mean volume loss was detected for the Silorane sample at about 6.4 μm^3 after 360.000 chewing cycles (87).

1.3.21.1 *Heliomolar®*

A Heliomolar composite was evaluated over a 3-year period by means of a controlled clinical trial (88). A total of 52 composite resin restorations and 52 amalgam alloy restorations were included in this trial. There were no significant differences in clinical performance between test and control materials, which both gave good service over the period of evaluation.

The components of Heliomolar are shown in Table 1.18.

Table 1.18 Components of Heliomolar (89).

Material	[%] per weight
Bis-GMA, Urethane dimethacrylate	19
Decanediol dimethacrylate	3
Triethylene glycol dimethacrylate	–
Highly dispersed SiO_2 prepolymer, YbF_3	77
Stabilizers, catalysts and pigments	<1

The fluoride-releasing capabilities of Heliomolar were measured (90). Heliomolar was proven to continuously release fluoride for over one year (89).

Also, a study was conducted using an artificial mouth model. This study showed that dentin and enamel can take up the fluoride released by Heliomolar (91).

1.3.21.2 *Voco Grandio®*

Voco Grandio is a universal light-cured 87% filled nano-hybrid composite for anterior and posterior fillings of all classes (92). It has a low shrinkage of only 1.57%, an excellent polishability, improved translucency, and a very high biocompatibility.

1.3.21.3 *Filtek® Silorane*

The main strategy to reduce shrinkage is to increase the filler load, thereby reducing the proportion of the resin. Since the shrinkage is caused by the resin, a lower proportion of the resin in a composite reduces the shrinkage. However, the shrinkage intrinsic to a methacrylate resin has remained a major challenge. Therefore,

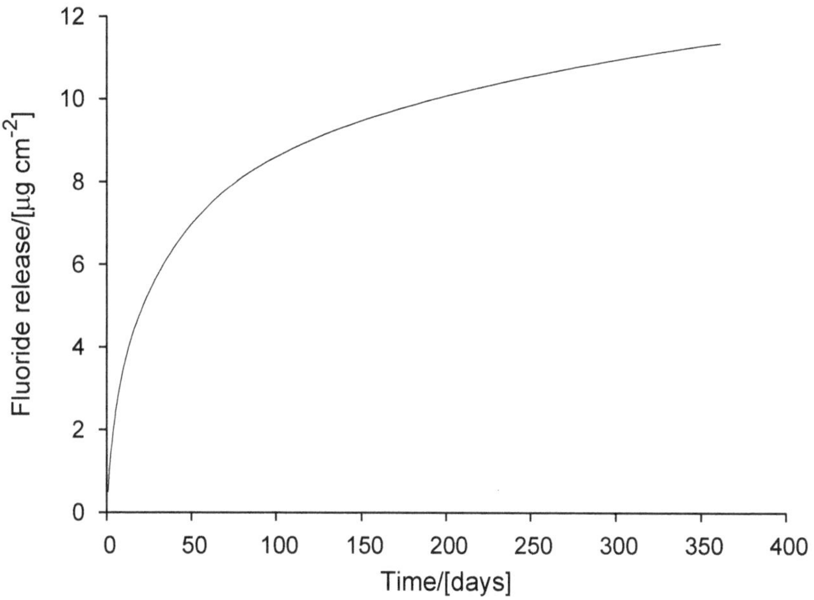

Figure 1.17 Fluoride release of Heliomolar.

exchanging the resin seems the most promising pathway to solve the shrinkage problem. The shrinkage of a Silorane and a methyl methacrylate-based resin is shown in Figure 1.18.

Due to its siloxane backbone, a silorane resin is more hydrophobic than conventional methacrylate resins. This results in a reduced water uptake.

Siloranes are a comparatively new class of compounds for use in dentistry. The name silorane originates from its chemical building blocks, siloxanes and oxiranes. The compound is shown in Figure 1.19.

The components of Silorane are shown in Table 1.19.

Initiator System. One component of the initiating system is camphorquinone, which matches the light spectrum of conventional dental polymerization light sources.

Further components of the initiating system are iodonium salts and electron donors, which generate the reactive cationic species

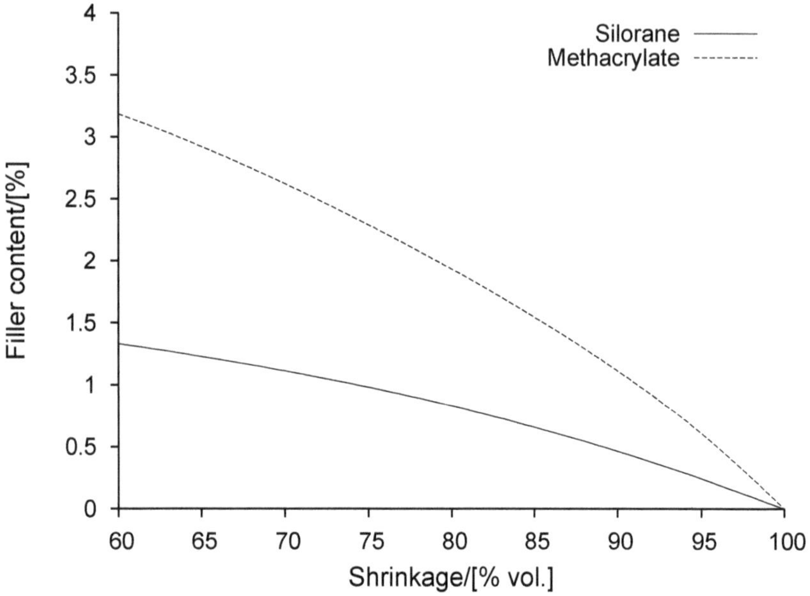

Figure 1.18 Shrinkage vs. filler content (3).

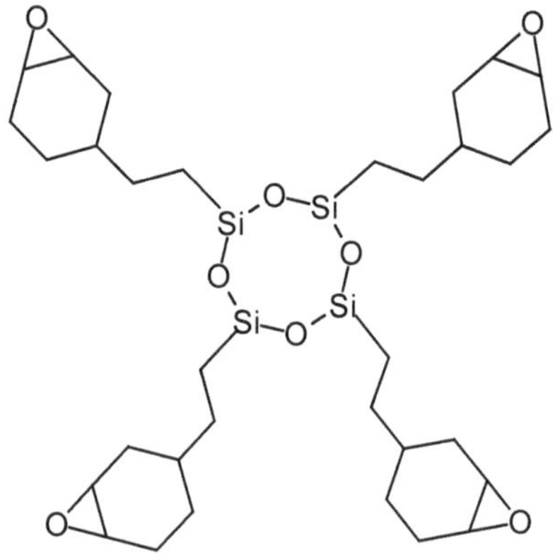

Figure 1.19 Silorane.

Table 1.19 Components in Filtek™ Silorane (3).

Material	[%] per weight
Silorane resin	23
Filler	76
Initiator	0.9
Stabilizer	0.13
Pigments	0.005

that can start the ring-opening polymerization. These compounds are shown in Figure 1.20.

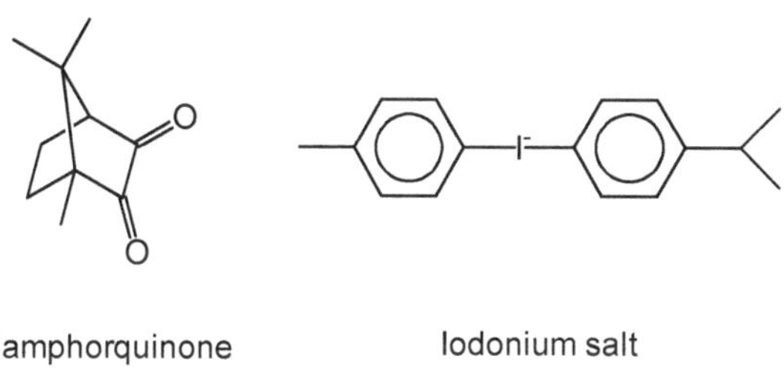

Figure 1.20 Components of the initiating system (3).

Adhesive Self-Etch Primer. Basically, self-adhesion is generated by acidic monomers that etch the dental substrates and thus create a retention pattern for micromechanical interlocking of the cured adhesive with the tooth. Furthermore, they provide chemical bonding to the calcium-containing hydroxyapatite of the mineralized tissue.

Most of the current self-etch adhesives contain phosphorylated methacrylates as acidic monomers. Some compositions contain carboxylic acid functionalized monomers, or a combination of both.

Here, the adhesive self-etch primer contains phosphorylated methacrylates, as well as Vitrebond™. This copolymer is from the group of resin-modified glass ionomers and has a carboxylic acid functionality (3).

1.3.21.4 *Leachable Residual Monomers and Additives*

The quality and quantity of leachable residual monomers, comonomers and additives eluted from various commercial dental composite resins after polymerization have been analyzed (93).

Polymerized specimens from four universal hybrid-type composite resins were eluted for 3 *d* with methanol or water. Then all extracts were analyzed by gas chromatography/mass spectrometry or liquid chromatography/mass spectrometry using a particle beam interface.

In all polymerized composite resin specimens, monomers and comonomers and various additives, as well as contaminants from manufacturing processes, could be identified. Almost every compound detected in the unpolymerized resins could also be identified in the methanol extracts, but only a few of them were found in the water extracts.

From these compounds, the comonomer triethylene glycol dimethacrylate was extracted in quantities higher than those reported to be cytotoxic in primary human oral fibroblast cultures (93).

It has been concluded that the extractable quantities of composite resin components should be minimized, either by reducing the mobility of leachable substances within the set material or by applying less water-soluble components.

In addition, all the ingredients of a dental composite should be declared by the manufacturers, in order to identify those substances in a product which may cause adverse side effects in patients and dental personnel (93).

1.4 Special Fabrication Methods

1.4.1 *Photoinitiator System for Hybrid Polymers*

The effect of the photoinitiator system on the properties of a dental material based on a hybrid polymer has been evaluated (94).

Two different dental composites were prepared, one containing a mixture of triethylene glycol dimethacrylate and bisphenol glycidyl dimethacrylate, TEGDMA/Bis-GMA, in a ratio of 50:50 and, another containing a mixture of bisphenol glycidyl dimethacrylate TEGDMA and an oligomeric inorganic precursor TEGDMA/p-MEMO in a ratio of 50:50. The composites were made with 70% of inorganic filler. Both composites were polymerized using Lucirin and camphorquinone as photoinitiators (94). These compounds are shown in Figure 1.21. Lucirin is also known as ethyl(2,4,6-trimethylbenzoyl)phenylphosphinate.

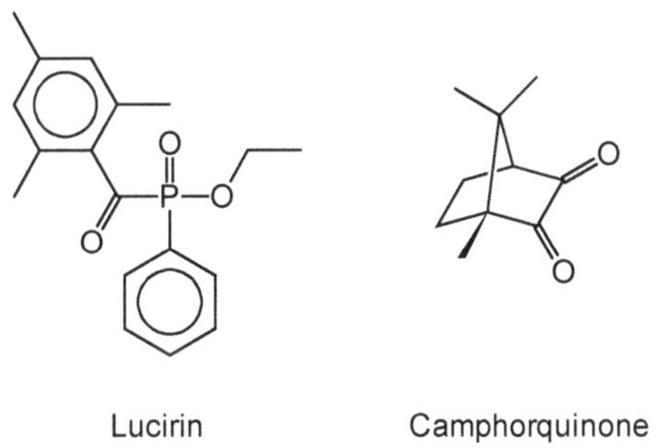

Figure 1.21 Lucirin and camphorquinone.

The flexural strength was evaluated with a universal test machine and the degree of conversion was measured using IR spectroscopy. To get the polymer shrinkage data a helium pycnometer was used.

It was found that Lucirin is the most suitable photoinitiator for dental composites containing hybrid polymers (94).

1.4.2 *Nanotechnology*

Some new developments in the field of nanotechnology applied to dentistry have been reviewed, focusing on the use of nanomaterials for improving the quality of oral care (95). Also, safety concerns regarding the use of dental nanomaterials have been discussed.

Several properties, such as morphological, antibacterial, mechanical, fluorescence, antitumoral, and remineralization and regeneration potential of polymeric, metallic and inorganic nano-based materials have been detailed, as well as their use as nanocluster fillers, in nanocomposites, mouthwashes, medicines, and biomimetic dental materials (95).

The fields of use of antimicrobial nanoparticles are collected in Table 1.20.

Table 1.20 Use of antimicrobial nanoparticles (95).

Material	Use
Silver nanoparticles	Root canal irrigation
Silver nanoparticles	Adhesives
Silver nanoparticles	Implants
Silver nanoparticles	Antibacterial formulations
Silver nanoparticles	Mouthwash
Zirconium oxide	Implants
Zirconium oxide	Adhesive cements
Zirconium oxide	Resin composites adhesives
Zirconium oxide	Antibacterial formulations
Zirconium oxide	Implants
Titanium dioxide	Resin composites adhesives
Titanium dioxide	Bleaching agents
Cuprous oxide	Resin composites adhesives
Chitosan nanoparticles	Resin composites adhesives
Chitosan nanoparticles	Implants
Quaternary ammonium compounds	Resin composites adhesives

Silver-based nanomaterials are effective against biofilms because they can attack multiple sites within the cell at a very low concentration of 0.5–1.0% to prevent bacterial growth (96–98).

The antibacterial activities against a panel of oral pathogenic bacteria and bacterial biofilms together with potential cytotoxic effects on human gingival fibroblasts of silver-based nanoparticles have been evaluated (99).

The highest antimicrobial activities at nontoxic concentrations were observed for the uncapped silver nanoparticles and the silver nanoparticles capped with lipoic acid (99). It was found that silver nanoparticles capped with lipoic acid and silver nanoparticles with poly(ethylene glycol) demonstrated a lower cytotoxicity than silver nanoparticles with tannic acid and uncapped materials.

The bactericidal effect of ZnO nanoparticles may be attributed to their ability to interact with the cell membrane of several bacterial species. Zn strongly binds to lipids and proteins, changing the osmotic balance and increasing membrane permeability (100–102). In addition, the nanoparticles increase the oxidative stress within the bacterial cell because of their ability to generate Zn^{2+} ions and reactive oxygen species, which can also inhibit the growth of planktonic bacteria.

Titanium dioxide-based nanoparticles undergo photocatalysis when they are exposed to a near-UV and UVA radiation, thus producing reactive oxygen species, mainly H_2O_2 and OH^-, which alter the osmotic equilibrium of bacteria. Furthermore, titanium dioxide-based nanoparticles can interfere with the phosphorylation reaction,thereby causing an oxidative cell death (103).

The antibacterial effect of chitosan, cf. Figure 1.22, may be attributed to its chemical structure, with the presence of deacetylated C_2 amino groups which become protonated and positively charged at a pH of less than 6.5. This leads to an increase in membrane permeability with a concomitant increase in the outward flow of ions and proteins from the microbial cell and to the inhibition of mRNA transcription and the alteration of protein translation, owing to binding of chitosan to the deoxyribonucleic acid of several microorganisms (95).

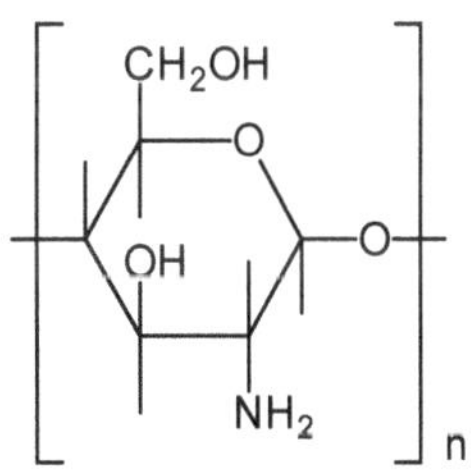

Figure 1.22 Chitosan.

A long-term antimicrobial PMMA-containing composition has been developed by incorporating mesoporous silica nanocarriers (104). A long-term antimicrobial effect was observed over 2 weeks due to the slow release of amphotericin B after loading it into the composition. Amphotericin B is shown in Figure 1.23.

1.4.3 *Extensive Dark Curing*

Radical-based photopolymerization is characterized by a rapid cessation of polymerization when the photocuring light source is extinguished. To maintain the active polymerization to completion,

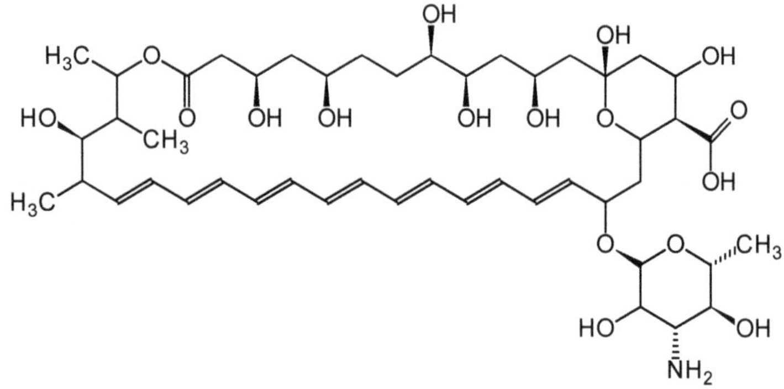

Figure 1.23 Amphotericin B.

continuous initiation is required because radical-based active centers have short propagating times due to highly efficient termination reactions (105).

In contrast, the cationic photopolymerization allows significant dark curing because of long active center lifetimes. But, compared with radical polymerization, there are few monomers compatible with the cationic curing process. This dark curing behavior is one of the distinct differences between cationic and free radical photopolymerization, and the prospect of removing this dark cure limitation from radical-based polymerization with conventional monomers is highly significant because effective dark curing would be useful to reduce processing times and lower initiator concentrations, as well as to achieve photocuring in shadow regions, enhancing the depths of cure and for photocuring pigmented or highly filled systems using radical polymerization (105).

There are only a few examples of free radically polymerizable monomers that exhibit considerable dark cure potential (106). These are (meth)acrylate monomers that are characterized by a number of unusual properties, such as hyperreactivity and significant formation of crosslinks from mono-vinyl polymerization, in addition to substantial dark cure potential with conventional photoinitiating systems.

A composition and a method for extensive dark curing from a visible-light-initiated controlled radical polymerization has been presented. Here, the radical active centers are not terminated, even when the light source is extinguished (105).

The three-component initiator system is a composition containing a photooxidizable photosensitizing agent, an electron donor, and an electron acceptor.

The monomers that can be used for such a composition are shown in Table 1.21 and in Figure 1.24. Other additives, such as photosensitizing agents, electron donors, and electron acceptors, are shown in Tables 1.22, 1.24, and 1.24, respectively. Photosensitizing agents, electron donors and electron acceptors, are also shown in Figures 1.25, 1.26, and 1.27, respectively.

Table 1.21 Monomers for dark curing (105).

Monomer
2-Hydroxyethyl acrylate
2-Hydroxyethyl methacrylate
Acrylamide
Methacrylamide
2,2-Bis[4-(2-hydroxy-3-methacryloxyprop-1-oxy)phenyl]propane
Urethane dimethacrylate
Glycerol monomethacrylate
1,3-Glycerol dimethacrylate
Tetrahydrofurfuryl methacrylate
1,6-Hexanediol diacrylate
1,6-Hexanediol dimethacrylate
Methacrylic acid
Triethylene glycol dimethacrylate
Styrene
Neodecyl vinyl ester

Importantly, the monomer should have at least one abstractable hydrogen. An example of such a formulation is presented in Table 1.25.

The dye, methylene blue, the electron donor, *N*-ethyldiisopropylamine, and the electron acceptor, diphenyl iodonium chloride, are dissolved completely upon addition to the monomer 2-hydroxyethyl methacrylate for 30 *min* using a vibration mixer at room temperature. After completely mixing the three-component initiators in a monomer, the formulation was directly used for a controlled radical dark polymerization (105).

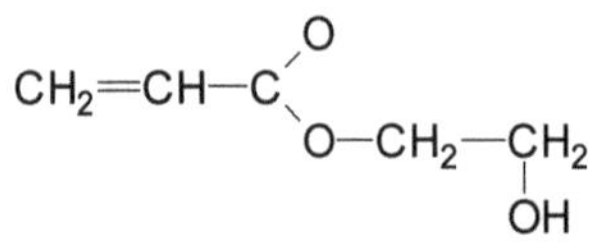

2-Hydroxyethyl acrylate

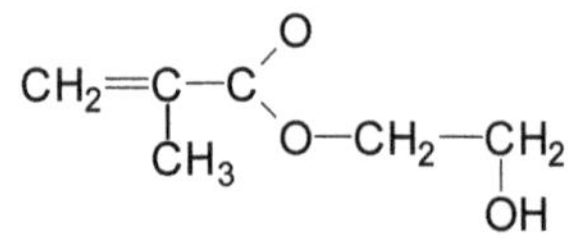

2-Hydroxyethyl methacrylate

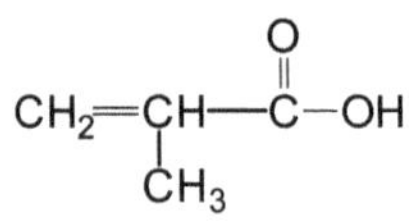

Metacrylic acid

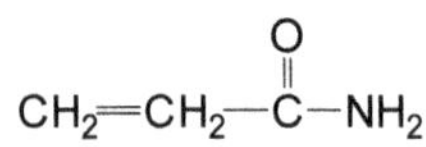

Acrylamide

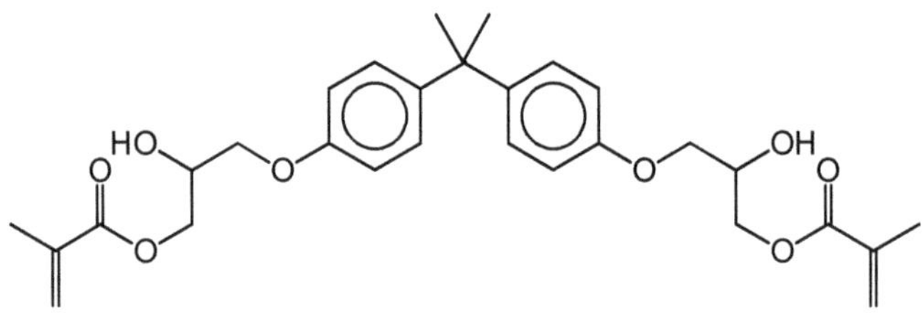

2,2-Bis[4-(2-hydroxy-3-methacryloxyprop-1-oxy)phenyl]propane

Figure 1.24 Monomers for dark curing.

Table 1.22 Photosensitizing agents (105).

Compound	Compound
Safranin O	Eosin Y disodium salt
Fluorescein sodium salt	Erythrosin B sodium salt
Acriflavine	Camphorquinone
Methylene blue	1-Phenyl-1,2-propanedione
Acridine orange	Resorufin
Resazurin	Phenosafranin
Rose bengal	Rhodamine B
Thioxanthen-9-one	

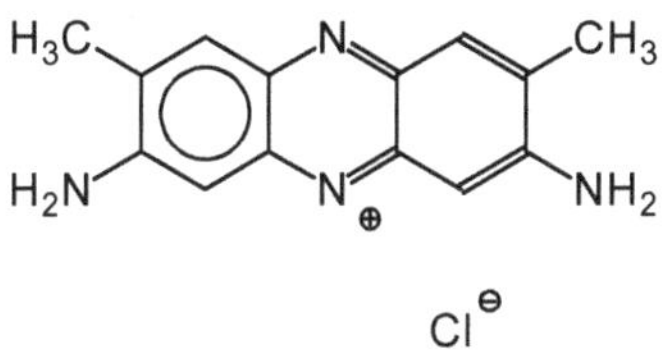

Safranin O

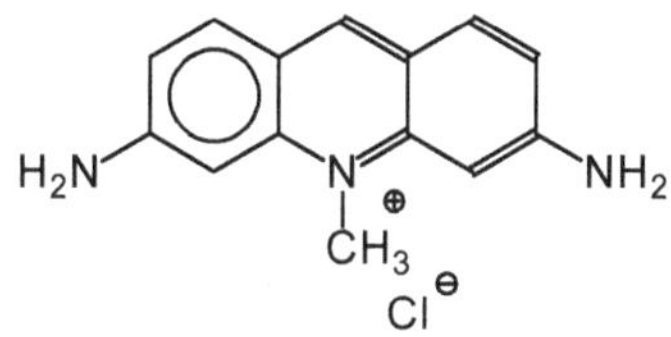

Acriflavine

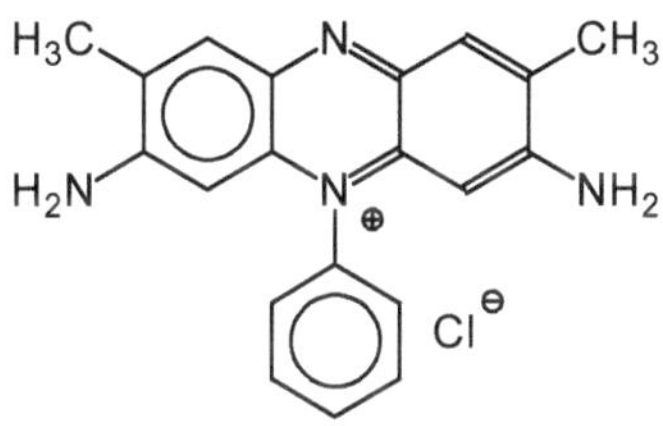

Phenosafranin

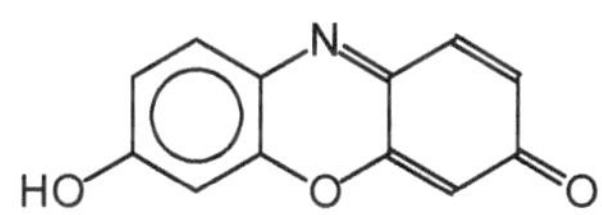

Resorufin

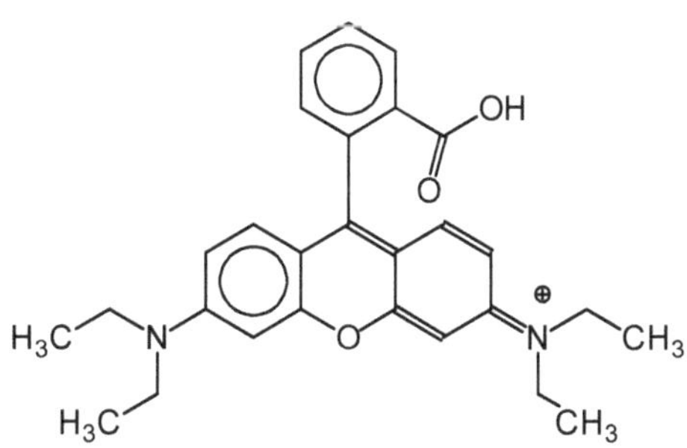

Rhodamine B

Figure 1.25 Photosensitizing agents.

Table 1.23 Electron donors (105).

Compound
Triethanolamine
4-Dimethylaminophenethyl alcohol
Ethyl 4-(dimethylamino)benzoate
Quinuclidine
N,N-Diisopropyl-3-pentylamine
N-Phenylglycine
N-Ethyldiisopropylamine
N-Methyldiethanolamine
Triethylamine
N,N-Dimethylacetamide
1,4-Diazabicyclo[2.2.2]octane
4-(Dimethylamino)benzaldehyde
N,N-Dimethylbenzylamine
4-*tert*-Butyl-*N,N*-dimethylaniline
9,10-Dimethylanthracene
N,N-Dimethylbenzylamine
1,2,2,6,6-Pentamethylpiperidine

Table 1.24 Electron acceptors (105).

Compound
Ferrocenium hexafluorophosphate
Diphenyl iodonium hexafluoroarsenate
4-[(2-Hydroxytetradecyl)oxyl]phenyl phenyliodonium hexafluoroantimonate
Diphenyl iodonium hexafluorophosphate
2,4,6-Tris(trifluoromethyl)-1,3,5-triazine
Diphenyl iodonium chloride
Diphenyl iodonium tetrafluoroborate

Table 1.25 Dark curable composition (105).

Ingredient	Weight/[*g*]
2-Hydroxyethyl methacrylate	5.3644
Methylene blue	0.0115
N-Ethyldiisopropylamine	0.1330
Diphenyl iodonium chloride	0.0653

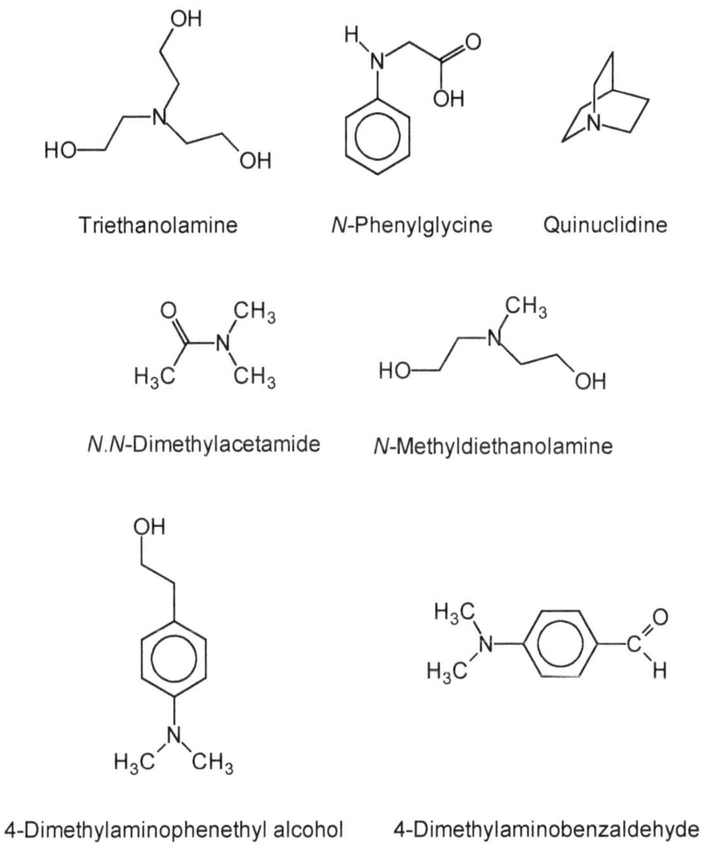

Figure 1.26 Electron donors.

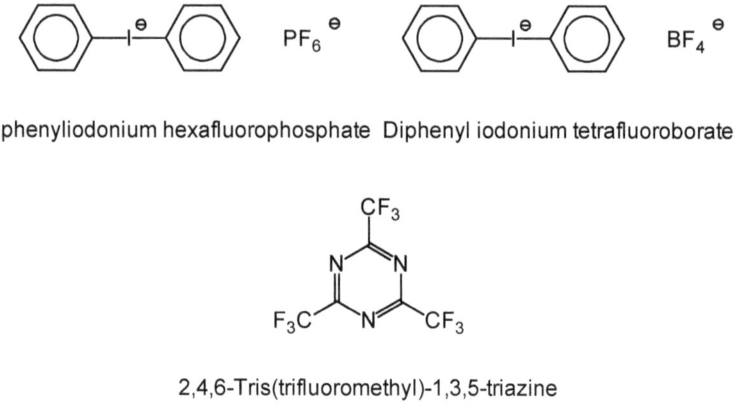

Figure 1.27 Electron acceptors.

For curing, visible light sources were used, a halogen lamp for low intensity experiments and modified dental curing light for high-intensity experiments. The monomer conversions as a function of the illumination time of the compositions shown in Table 1.25 are shown in Table 1.26 and in Figure 1.28.

Table 1.26 Monomer conversions as a function of illumination time (105).

Illumination time/[s]	Initial Conversion/[%]	Final Conversion/[%]
76	43.3	80.0
65	23.8	74.4
50	17.2	67.4
30	8.6	52.5
20	6.1	48.1
10	1.3	44.6
0	0.0	0.0

1.4.4 *Oral Implantology*

The fabrication of temporary restorations is an essential part of dentistry and in particular of oral implantology (107). Poly(methyl methacrylate) has been used for a long time as a material for tem-

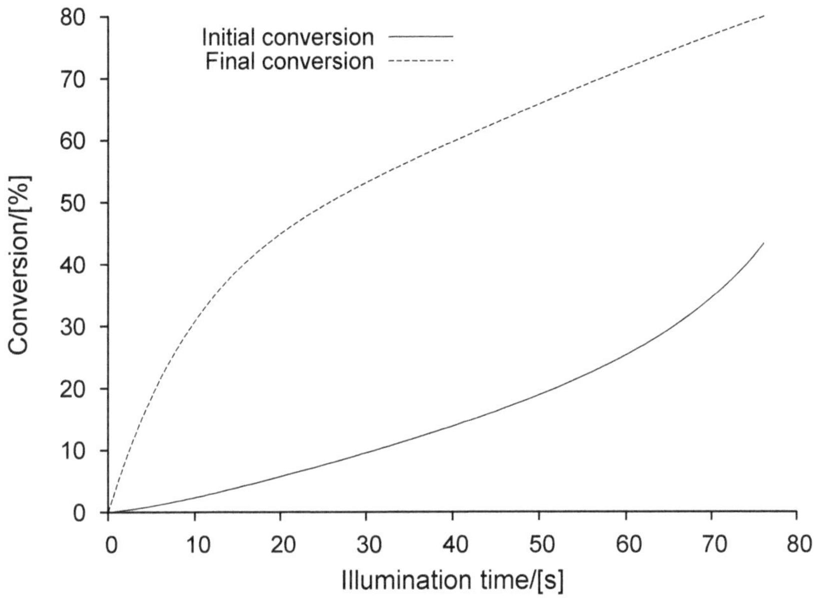

Figure 1.28 Monomer conversion as a function of illumination time (105).

porary restorations. Many improvements have been reported to increase its properties, especially fracture resistance.

Crosslinking and computer-assisted design and computer-assisted machining permit the fabrication of long-term temporary restorations with a flexural strength high enough and good modulus of elasticity that do not fracture easily under functional loads.

The fabrication of a PMMA-resin CAD/CAM temporary implant-supported crown follows the workflow (107):

- Evaluation of the implant site,
- Impression with poly(vinylsiloxane) in open tray,
- Pouring the model and preparation for scanning,
- Design and CAD/CAM fabrication of crown from a monolithic block Telio CAD, utilizing a three-axis milling unit,
- Verification of adaptation of crown on implant abutment, and
- Placement in the oral cavity.

1.4.4.1 *Bisphosphonates*

A systemic literature review has been presented to evaluate the potential capacity of the topical application of bisphosphonates to preserve/enhance alveolar bone in oral implantology (108).

Bisphosphonates are a group of drugs commonly used for the treatment of various bone diseases, including osteoporosis, malignant hypercalcemia, multiple myeloma, or Paget's disease (108).

Two groups of bisphosphonates are available, with different mechanisms of action: Amino and non-amino-bisphosphonates. Non-amino-bisphosphonates, such as clodronate and etidronate, inhibit the bone resorption primarily by inducing osteoclast apoptosis through the formation of intracellular metabolites in osteoclasts. Amino-bisphosphonates, such as pamidronate, alendronate or zoledronate, offer a greater potency through the addition of a primary amino-nitrogenated base ($-NH_2$) (109, 110). Some of the corresponding acid compounds are shown in Figure 1.29.

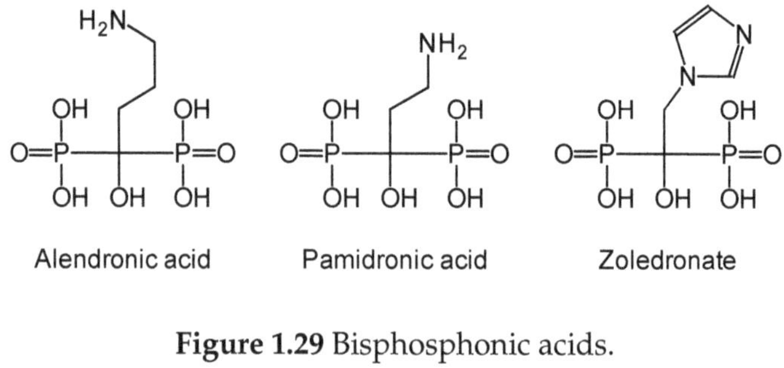

Figure 1.29 Bisphosphonic acids.

These act by inhibiting farnesyl diphosphate synthase, a key enzyme in the mevalonate pathway (111).

Mevalonic acid is a precursor in the biosynthetic pathway, which is known as the mevalonate pathway that produces terpenes and steroids. Mevalonic acid, cf. Figure 1.30, is the primary precursor of isopentenyl pyrophosphate that is actually the basis for all terpenoids (112, 113).

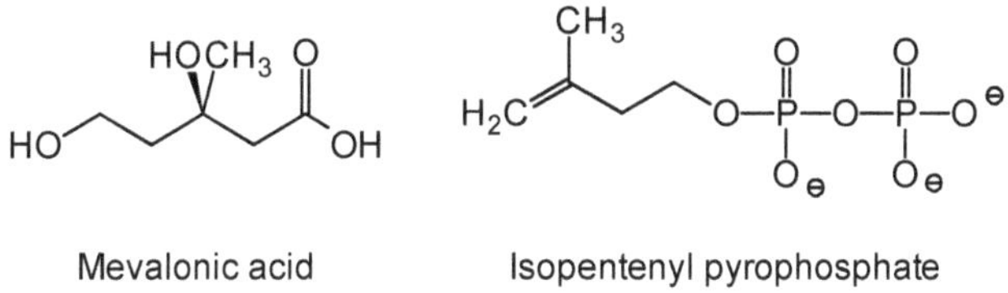

Figure 1.30 Mevalonic acid and isopentenyl pyrophosphate.

As a consequence of their high affinity for Ca^{2+} ions, bisphosphonates are rapidly cleared from circulation and target hydroxyapatite bone mineral surfaces *in-vivo* at sites of active bone remodeling. Several experimental studies have demonstrated that these drugs reduce bone resorption by inhibiting the activity of mature osteoclasts and promoting their apoptosis (114, 115).

They also inhibit the formation and recruitment of new osteoclasts, suppressing the osteoclasts' multinucleated cells during the osteoclast differentiation process. In addition, recent experimental studies have demonstrated that some bisphosphonates enhance osteoblast differentiation and activity. For example, alendronate and clodronate seem to act directly on these cells, stimulating differentiation, proliferation, and bone formation/mineralization.

Traditionally, bisphosphonates have been administrated both intravenously and orally. In a beagle dog study, is was observed that the systemic administration of bisphosphonates could prevent the alveolar bone destruction associated with periodontal disease (116). However, in recent years a worrying correlation has emerged between osteonecrosis of the jaw and the systemic administration of bisphosphonates (117–120).

Because of these potential risks of intravenous bisphosphonate administration, other methods have been proposed (108). Yaffe and coworkers could demonstrate that the topical application of bisphosphonates minimizes the bone resorption following a mucoperiostial flap surgery (121–123).

In another study (124), peri-implantitis was induced by ligation around the abutments, 6 months after placement of a fixture.

Pamidronate in an amount of 0.6 $mg\,kg^{-1}$ was injected intramuscularly every 3 d into each of 5 dogs. Another 5 dogs served as the control group and were injected with saline only. Peripheral blood and urine samples were collected every week up to 12 weeks after the placement of the ligature. It was observed that topical bisphosphonates inhibited the progression of alveolar bone resorption in peri-implantitis (124).

1.4.5 *Dental Inlays*

1.4.5.1 Synthetic Poly(urethane) Composite

Osteoimplants can be used in a variety of orthopedic, neurosurgical, dental, and oral and maxillofacial surgical procedures such as the repair of simple and compound fractures and non-unions, external, and internal fixations, joint reconstructions such as arthrodesis, general arthroplasty, deficit filling, disectomy, laminectomy, anterior cerival and thoracic operations (125).

Injectable, functionally weight-bearing biomaterials that possess both initial mechanical strength comparable to host bone and maintain their initial strength while actively remodeling to form new bone would transform clinical management of a number of orthopedic conditions. Functionally weight-bearing biomaterials for treatment of bone defects ideally possess five qualities (125):

1. Biocompatibility of the material and its breakdown products,
2. Injectability to enable less invasive application and fill irregularly shaped defects,
3. Weight-bearing properties with strength comparable to that of healthy host bone at the defect site,
4. Support of rapid cellular infiltration and remodeling at a rate that does not inhibit bone repair, and
5. Delivery of biologics with proper release kinetics to accelerate bone formation and remodeling.

Poly(urethane) resins (PURs) are a useful class of biomaterials due to the fact that they can be injectable or moldable as a reactive liquid that subsequently cures to form a porous composite. These materials also have tunable degradation rates, which are shown to be highly dependent on the choice of polyol and isocyanate components (126).

Also, PURs have tunable mechanical properties, which can also be enhanced with the addition of osteoconductive matrix or other components. The materials exhibit elastomeric rather than brittle mechanical properties. PURs can be made by reacting together the components of a two-component composition, one of which includes a polyisocyanate while the other includes a component having two or more hydroxyl groups, i.e., polyols to react with the polyisocyanate (125).

1.4.5.2 *Living Polymers*

Living polymerization was discovered in the 1950s. It was first demonstrated by Michael Szwarc in 1956 in the anionic polymerization of styrene with an alkali metal/naphthalene system in tetrahydrofuran (THF). Szwarc found that, after addition of monomer to the initiator system, an increase in viscosity would eventually cease but that the viscosity would start to increase again after addition of a new amount of monomer (en.wikipedia.org/wiki/Living polymerization). Since then, living polymerization has evolved and, in many laboratories around the world, conditions for obtaining such polymerization were discovered for various other types of anionic, cationic, ring-opening and free radical systems. Protection of the living end of a polymer from termination has been accomplished by complexation or by steric hindrance and by an appropriate choice of reagents and solvents (127).

A kit for providing a polymerizable resin system has been described, which comprises a first Part A and a second Part B (128). Part A and Part B upon mixing provides a working period of intermediate stage polymerization in which the mixture obtains a desired cohesiveness for a predetermined period of time.

The first Part A is an acid and the second Part B is an organic compound that is water-soluble or partially water-soluble and that, in the presence of the acid, initiates curing of polymerizable monomer and/or resin that is present in Part A, Part B, or both (128).

References

1. J.L. Ferracane, *Materials in Dentistry: Principles and Applications*, Lippincott Williams & Wilkins, Philadelphia, 2nd edition, 2001.

2. J. Powers and J. Wataha, *Dental Materials: Properties and Manipulation*, Elsevier Health Sciences, London, 10th edition, 2014.
3. N. N., Filtek™ Silorane, Scienitific documentation, 3M ESPE AG, ESPE Platz 82229 Seefeld, Germany, 2007. electronic: http://multimedia.3m.com/mws/media/598061O/filtek-silorane-tpp-ebu.pdf.
4. K.L.V. Landuyt, J. Snauwaert, J.D. Munck, M. Peumans, Y. Yoshida, A. Poitevin, E. Coutinho, K. Suzuki, P. Lambrechts, and B.V. Meerbeek, *Biomaterials*, Vol. 28, p. 3757, 2007.
5. Technical Committee: ISO/TC 106/SC 2, Dental materials – determination of colour stability, ISO Standard ISO 7491, International Organization for Standardization, Geneva, Switzerland, 2000.
6. Technical Committee: ISO/TC 106/SC 2, Dentistry — polymer-based crown and bridge materials, ISO Standard ISO 10477, International Organization for Standardization, Geneva, Switzerland, 2004.
7. Technical Committee: ISO/TC 106/SC 2, Dentistry – soft lining materials for removable dentures – part 1: Materials for short-term use, ISO Standard ISO 10139-1, International Organization for Standardization, Geneva, Switzerland, 2005.
8. Technical Committee: ISO/TC 106/SC 2, Dentistry – brazing materials, ISO Standard ISO 9333, International Organization for Standardization, Geneva, Switzerland, 2006.
9. Technical Committee: ISO/TC 106/SC 2, Dentistry – polymer-based restorative materials, ISO Standard ISO 4049, International Organization for Standardization, Geneva, Switzerland, 2009.
10. Technical Committee: ISO/TC 106/SC 2, Dentistry – compatibility testing – part 1: Metal-ceramic systems, ISO Standard ISO 9693-1, International Organization for Standardization, Geneva, Switzerland, 2012.
11. ASTM International, Standard test method for objective measurement of gingival color using digital still cameras, ASTM Standard ASTM E2545-07, ASTM International, West Conshohocken, PA, 2012.
12. Technical Committee: ISO/TC 106/SC 2, Dentistry – elastomeric impression materials, ISO Standard ISO 4823, International Organization for Standardization, Geneva, Switzerland, 2015.
13. Technical Committee: ISO/TC 106/SC 2, Dentistry – ceramic materials, ISO Standard ISO 6872, International Organization for Standardization, Geneva, Switzerland, 2015.
14. ASTM International, Standard practice for managing sustainability in dentistry, ASTM Standard ASTM E3014-15, ASTM International, West Conshohocken, PA, 2015.
15. Technical Committee: ISO/TC 106/SC 2, Dentistry – compatibility testing – part 2: Ceramic-ceramic systems, ISO Standard ISO 9693-2,

International Organization for Standardization, Geneva, Switzerland, 2016.
16. Technical Committee: ISO/TC 106/SC 2, Dentistry – part 2: Materials for long-term use, ISO Standard ISO 10139-2, International Organization for Standardization, Geneva, Switzerland, 2016.
17. Technical Committee: ISO/TC 106/SC 1, Dentistry – polymer-based luting materials containing adhesive components, ISO Standard ISO/TS 16506, International Organization for Standardization, Geneva, Switzerland, 2017.
18. ISO Standards Catalogue, Dental materials, 2017. electronic: https://www.iso.org/ics/11.060.10/x/.
19. D.J. Gardner, *Theories and mechanisms of adhesion in the pharmaceutical, biomedical and dental fields* in K.L. Mittal and F.M. Etzler, eds., *Adhesion in Pharmaceutical, Biomedical, and Dental Fields*. John Wiley & Sons, 2017.
20. A. Vinagre and J. Ramos, *Adhesion in restorative dentistry* in A. Rudawska, ed., *Materials Science: Adhesives – Applications and Properties*, chapter 3. InTech, Rijeka, Croatia, 2016.
21. L. Alkadi and N.D. Ruse, *The Journal of Prosthetic Dentistry*, Vol. 116, p. 591, 2016.
22. A. Tadin, L. Gavic, and N. Galic, *Biocompatibility of dental adhesives* in A. Rudawska, ed., *Adhesives - Applications and Properties*, chapter 2. InTech, Rijeka, Croatia, 2016.
23. A. Doyle and J.B. Griffiths, eds., *Cell and Tissue Culture: Laboratory Procedures in Biotechnology*, Wiley, Chichester New York, 1998.
24. Wikipedia, Trypan blue — wikipedia, the free encyclopedia, 2017. [Online; accessed 4-September-2017].
25. A. Tadin, N. Galic, D. Zeljezic, B.M. Vitasovic, D. Marovic, and I. Kovacic, *Journal of Dental Sciences*, Vol. 8, p. 37, 2013.
26. D. Marovic, A. Tadin, M. Mladinic, D. Juric-Kacunic, and N. Galic, *American Journal of Dentistry*, Vol. 27, p. 35, February 2014.
27. J.-F. Roulet, *Degradation of Dental Polymers*, Karger, Basel New York, 1987.
28. J.P. Santerre, L. Shajii, and B.W. Leung, *Critical Reviews in Oral Biology & Medicine*, Vol. 12, p. 136, 2001.
29. R.L. Bowen and W.A. Marjenhoff, *Advances in Dental Research*, Vol. 6, p. 44, 1992.
30. B. Sandner and R. Schreiber, *Die Makromolekulare Chemie*, Vol. 193, p. 2763, 1992.
31. M. Jaymand, M. Lotfi, J. Barar, and S. Kimyai, *Research on Chemical Intermediates*, Vol. 43, p. 5707, Oct 2017.
32. T.T. Tauböck, A.J. Feilzer, W. Buchalla, C.J. Kleverlaan, I. Krejci, and T. Attin, *European Journal of Oral Sciences*, Vol. 122, p. 293, 2014.

33. Wikipedia, Knoop hardness test — wikipedia, the free encyclopedia, 2017. [Online; accessed 6-September-2017].
34. R. Belli, E. Geinzer, A. Muschweck, A. Petschelt, and U. Lohbauer, *Dental Materials*, Vol. 30, p. 424, 2014.
35. A. Tahayeri, M. Morgan, A.P. Fugolin, D. Bompolaki, A. Athirasala, C.S. Pfeifer, J.L. Ferracane, and L.E. Bertassoni, *Dental Materials*, 2017.
36. Nextdent, NextDent C&B. Biocompatible printable material for bridges and crowns, electronic: https://nextdent.com/products/cb/, 2017.
37. M.B. Güngör, S.K. Nemli, B.T. Bal, and A. Doğan, *Acta Odontologica Turcica*, Vol. 33, p. 80, 2016.
38. Ivoclar Vivadent, Telio CAD, electronic: http://www.ivoclarvivadent.com/en/p/all/products/temporary-materials/telio-product-system/telio-cad, 2017.
39. C. Toriyabe and H. Akizumi, Dental filling and restorative material kit, US Patent 9 687 421, assigned to Tokuyama Dental Corporation (Tokyo, JP), June 27, 2017.
40. H. Fong, S.H. Dickens, and G.M. Flaim, *Dental Materials*, Vol. 21, p. 520, 2005.
41. L.V.J. Lassila, P.K. Vallittu, S. Garoushi, and K. Airola, Fiber-reinforced composites and method for the manufacture thereof, US Patent 8 785 513, assigned to Stick Tech. Oy (Turku, FI), July 22, 2014.
42. N. Moszner, J. Angermann, V. Rheinberger, D. Voser, K. Vogel, and S. Klapdohr, Polymer-coated glass filler for use in dental materials, US Patent 7 932 304, assigned to Ivoclar Vivadent AG (Schaan, LI), April 26, 2011.
43. N. Moszner and U. Salz, *Macromolecular Materials and Engineering*, Vol. 292, p. 245, 2007.
44. T. Nakatuka, Y. Yasuda, K. Kimoto, M. Mizuno, and N. Negoro, Dental fillers, US Patent 6 620 861, assigned to Kabushiki Kaisha Shofu (Kyoto, JP), September 16, 2003.
45. S.K. Sidhu and J.W. Nicholson, *Journal of Functional Biomaterials*, Vol. 7, p. 16, 2016.
46. Technical Committee: ISO/TC 106/SC 1, Dental water based cements: Part 1: Powder/liquid acid-base cements, ISO Standard ISO 9917-1, International Organization for Standardization, Geneva, Switzerland, 2007.
47. J.E. Klee, O. Elsner, S. Pohle, H. Ritter, and D. Bernert, Dental composition, US Patent 9 682 019, assigned to Dentsply DeTry GmbH (Constance, DE), June 20, 2017.
48. D. Khvostenko, J. Mitchell, T. Hilton, J. Ferracane, and J. Kruzic, *Dental Materials*, Vol. 29, p. 1139, 2013.
49. M.H. Alanzi, A.M. Almadani, and A. Alazemi, *Life Science Journal*, Vol. 14, 2017.

50. H. Kattan, X. Chatzistavrou, J. Boynton, J. Dennison, P. Yaman, and P. Papagerakis, *Materials*, Vol. 8, p. 4668, 2015.
51. H.-K. Hyun and J.L. Ferracane, *Dental Materials*, Vol. 32, p. 1144, 2016.
52. S. Heid, P.R. Stoessel, T.T. Tauböck, W.J. Stark, M. Zehnder, and D. Mohn, *Biomedical Glasses*, Vol. 2, p. 29, 2016.
53. Wikipedia, Bioglass — wikipedia, the free encyclopedia, 2017. [Accessed 29-November-2017].
54. A. Pegoretti and A. Dorigato, *Fatigue behaviour of biocomposites* in L. Ambrosio, ed., *Biomedical Composites (Second Edition)*, Woodhead Publishing Series in Biomaterials, pp. 431–478. Woodhead Publishing, 2nd edition, 2017.
55. M. Montazerian and E.D. Zanotto, *Restorative dental glass-ceramics: Current status and trends* in G. Kaur, ed., *Clinical Applications of Biomaterials: State-of-the-Art Progress, Trends, and Novel Approaches*, chapter 9, pp. 313–336. Springer International Publishing, Cham, 2017.
56. I.D. Sideridou and E.C. Vouvoudi, *Dental composites: Dimethacrylate-based* in M. Mishra, ed., *Encyclopedia of Biomedical Polymers and Polymeric Biomaterials*, pp. 2464–2470. Taylor and Francis, New York, 2015.
57. J.A. Orlowski, D.V. Butler, and A. Chin, Dental prosthetics comprising curable acrylate polymer compositions and methods of their use, US Patent 9 028 254, assigned to Scientific Pharmaceuticals, Inc. (Pomona, CA), May 12, 2015.
58. S.A. McAdams, W.D. Block, S.J. McCarthy, and K.W. Gregory, Compositions, assemblies, and methods applied during or after a dental procedure to ameliorate fluid loss and/or promote healing, using a hydrophilic polymer sponge structure such as chitosan, US Patent 7 897 832, assigned to HemCon Medical Technologies, Inc. (Portland, OR), March 1, 2011.
59. S.A. McAdams, W.D. Block, S.J. McCarthy, K.W. Gregory, and H. Oien, Compositions, assemblies, and methods applied during or after a dental procedure to ameliorate fluid loss and/or promote healing, using a hydrophilic polymer sponge structure such as chitosan, US Patent 9 004 918, assigned to HemCon Medical Technologies, Inc. (Portland, OR), April 14, 2015.
60. B.J. Sun, Hybrid polymer network compositions for use in dental applications, US Patent 8 765 836, assigned to DENTSPLY International Inc. (York, PA), July 1, 2014.
61. T. Buruiana, V. Melinte, H. Aldea, I.M. Pelin, and E.C. Buruiana, *Materials Science and Engineering: C*, Vol. 62, p. 96, 2016.
62. S.-B. Lee, C. González-Cabezas, K.-M. Kim, K.-N. Kim, and K. Kuroda, *Biomacromolecules*, Vol. 16, p. 2265, 2015.

63. A.S. Abuelyaman, A.R. Fornof, L.R. Krepski, and W.H. Moser, Dental composition containing high refractive index monomers, US Patent 9 675 529, assigned to 3M Innovative Properties Company (St. Paul, MN), June 13, 2017.
64. M.A.S. Melo, M.D. Weir, F. Fakhari, L. Cheng, K. Zhang, F. Li, X. Zhou, Y. Bai, and H.H.K. Xu, *Antibacterial polymers for dental adhesives and composites* in *Adhesion in Pharmaceutical, Biomedical and Dental Fields*, chapter 13, pp. 299–330. John Wiley & Sons, Inc., 2017.
65. J.D.B. Featherstone, *The Journal of the American Dental Association*, Vol. 131, p. 887, 2000.
66. H. Xu and M. Weir, Antibacterial monomers, antibacterial resins and dental composites comprising the antibacterial resins, US Patent Application 20 150 299 345, assigned to University of Maryland, Baltimore, (Baltimore MD), October 22, 2015.
67. J.D.B. Featherstone, *Journal of Dental Research*, Vol. 83, p. 39, 2004.
68. P. Totiam, C. González-Cabezas, M.R. Fontana, and D.T. Zero, *Caries Research*, Vol. 41, p. 467, 2007.
69. M.S. Cenci, T. Pereira-Cenci, J.A. Cury, and J.M. ten Cate, *Caries Research*, Vol. 43, p. 97, 2009.
70. P. Stoodley, J. Wefel, A. Gieseke, D. deBeer, and C. von Ohle, *The Journal of the American Dental Association*, Vol. 139, p. 1182, 2008.
71. A.S. Melo, M.D. Weir, F. Fakhari, L. Cheng, K. Zhang, F. Li, X. Zhou, Y. Bai, and H.H.K. Xu, *Antibacterial polymers for dental adhesives and composites* in K.L. Mittal and F.M. Etzler, eds., *Adhesion in Pharmaceutical, Biomedical and Dental Fields*, chapter 13, pp. 301–330. Scrivener Publishing John Wiley & Sons, Beverly, MA Hoboken, NJ, 2017.
72. S. Imazato, S. Ma, J. hua Chen, and H.H. Xu, *Dental Materials*, Vol. 30, p. 97, 2014.
73. A.R. Cocco, W.L. de Oliveira da Rosa, A.F. da Silva, R.G. Lund, and E. Piva, *Dental Materials*, Vol. 31, p. 1345, 2015.
74. M.A. Melo, S. Orrego, M.D. Weir, H.H.K. Xu, and D.D. Arola, *ACS Applied Materials & Interfaces*, Vol. 8, p. 11779, 2016.
75. S. Imazato, T. Imai, R.R.B. Russell, M. Torii, and S. Ebisu, *Journal of Biomedical Materials Research*, Vol. 39, p. 511, 1998.
76. L. Cheng, M.D. Weir, K. Zhang, D.D. Arola, X. Zhou, and H.H. Xu, *Journal of Dentistry*, Vol. 41, p. 345, 2013.
77. W. Zhang, X. juan Luo, L. na Niu, S. ying Liu, W. chun Zhu, J. Epasinghe, L. Chen, G. hua Li, C. Huang, J. Mao, D.H. Pashley, and F.R. Tay, *Journal of Dentistry*, Vol. 42, p. 1078, 2014.
78. N. Menschutkin, *Zeitschrift für Physikalische Chemie*, Vol. 5, p. 589, 1890.
79. J.M. Antonucci, D.N. Zeiger, K. Tang, S. Lin-Gibson, B.O. Fowler, and N.J. Lin, *Dental Materials*, Vol. 28, p. 219, 2012.

80. U. Ashraf, K.H. Tsoi, E.H.N. Pow, and J.P. Matinlinna, *Journal of Dental Research*, 2017.
81. A. Fugolin, A. Dobson, W. Mbiya, J. Ferracane, and C.S. Pfeifer, *Dental Materials*, Vol. 33, p. e34, 2017.
82. H. Xu and M. Weir, Nanostructured antibacterial and remineralizing dental bonding agents and dental bonding systems, US Patent 9 554 971, assigned to University of Maryland, Baltimore (Baltimore, MD), January 31, 2017.
83. H. Xu, M.D. Weir, and L. Zhang, Rechargeable calcium phosphate-containing dental materials, US Patent Application 20 160 228 335, assigned to University of Maryland, Baltimore, (Baltimore MD), August 11, 2016.
84. L. Zhang, M.D. Weir, L.C. Chow, J.M. Antonucci, J. Chen, and H.H.K. Xu, *Dental Materials*, Vol. 32, p. 285, 2016.
85. J. Hieda, M. Niinomi, M. Nakai, and K. Cho, *Dental materials* in R. Pignatello, ed., *Materials Science Biomaterials: Advances in Biomaterials Science and Biomedical Applications*, chapter 20. InTech, Rijeka, Croatia, 2013.
86. P.A. Sagel, L.N. Nguyen, and A.M. Tunis, Tooth desensitizing oral care compositions, devices, and methods, US Patent 9 717 665, assigned to The Procter & Gamble Company (Cincinnati, OH), August 1, 2017.
87. E. Yilmaz, R. Sadeler, Z. Duymus, and M. Ocal, *Biomedical and Biotechnology Research Journal*, Vol. 1, p. 25, 2017.
88. P.J. Knibbs and E.R. Smart, *Journal of Oral Rehabilitation*, Vol. 19, p. 231, 1992.
89. U. Lendenmann, The Heliomolar® Family: Heliomolar® Heliomolar® Flow Heliomolar® HB, Scienitific documentation, Ivoclar Vivadent, Bendererstrasse 2, FL - 9494 Schaan, Liechtenstein, 2001. electronic: http://downloads.ivoclarvivadent.com/zoolu-website/media/document/954/Heliomolar+Family?_ga=2.194815213.1300279295.1504166264-1952391403.1504166264.
90. J. Arends, J. Ruben, and A.G. Dijkman, *Quintessence International*, Vol. 21, 1990.
91. J. Arends and Y. Van der Zee, *Quintessence International*, Vol. 21, 1990.
92. VOCO GmbH, Grandio, http://www.voco.com/us/product/grandio/index.html, 2017.
93. W. Spahl, H. Budzikiewicz, and W. Geurtsen, *Journal of Dentistry*, Vol. 26, p. 137, 1998.
94. S.A. de Souza Venter, G.C. Martim, A. dos Santos, T.R. Detomini, E. Radovanovic, and E.M. Girotto, *Brazilian Dental Science*, Vol. 19, p. 96, 2016.
95. G.C. Padovani, V.P. Feitosa, S. Sauro, F.R. Tay, G. Durán, A.J. Paula, and N. Durán, *Trends in Biotechnology*, Vol. 33, p. 621, 2015.

96. P.D. Marcato and N. Durán, *Biogenic silver nanoparticles: Application in medicines and textiles and their health implications* in M. Rai and N. Duran, eds., *Metal Nanoparticles in Microbiology*, pp. 249–267. Springer Berlin Heidelberg, Berlin, Heidelberg, 2011.
97. N. Durán, P.D. Marcato, M. Durán, A. Yadav, A. Gade, and M. Rai, *Applied Microbiology and Biotechnology*, Vol. 90, p. 1609, Jun 2011.
98. R. de Lima, A.B. Seabra, and N. Durán, *Journal of Applied Toxicology*, Vol. 32, p. 867, 2012.
99. K. Niska, N. Knap, A. Kedzia, M. Jaskiewicz, W. Kamysz, and I. Inkielewicz-Stepniak, *International Journal of Medical Sciences*, Vol. 13, p. 772, September 2016.
100. Z. Huang, X. Zheng, D. Yan, G. Yin, X. Liao, Y. Kang, Y. Yao, D. Huang, and B. Hao, *Langmuir*, Vol. 24, p. 4140, 2008.
101. A.J. Huh and Y.J. Kwon, *Journal of Controlled Release*, Vol. 156, p. 128, 2011.
102. M.J. Hajipour, K.M. Fromm, A.A. Ashkarran, D.J. de Aberasturi, I.R. de Larramendi, T. Rojo, V. Serpooshan, W.J. Parak, and M. Mahmoudi, *Trends in Biotechnology*, Vol. 30, p. 499, 2012.
103. K. Blecher, A. Nasir, and A. Friedman, *Virulence*, Vol. 2, p. 395, 2011.
104. J.-H. Lee, A. El-Fiqi, J.-K. Jo, D.-A. Kim, S.-C. Kim, S.-K. Jun, H.-W. Kim, and H.-H. Lee, *Dental Materials*, Vol. 32, p. 1564, 2016.
105. J.W. Stansbury and D. Kim, Methods for extensive dark curing based on visible-light initiated, controlled radical polymerization, US Patent 8 883 948, assigned to The Regents of the University of Colorado (Denver, CO), November 11, 2014.
106. H. Kilambi, S.K. Reddy, L. Schneidewind, J.W. Stansbury, and C.N. Bowman, *Polymer*, Vol. 48, p. 2014 , 2007.
107. C.F. Andreescu, D.L. Ghergic, O. Botoaca, H.M. Barbu, I.S.C. Mitariu, and D.N. Patroi, *Materiale Plastice*, Vol. 54, p. 32, 2017.
108. N. Lozano-Carrascal, O. Salomó-Coll, F. Hernández-Alfaro, S.-A. Gehrke, J. Gargallo-Albiol, and J.-L. Calvo-Guirado, *Medicina Oral, Patología Oral y Cirugía Bucal*, Vol. 22, p. e512, 2017.
109. M.J. Rogers, *Current Pharmaceutical Design*, Vol. 9, p. 2643, 2003.
110. D. Berardi, T. Carlesi, F. Rossi, M. Calderini, R. Volpi, and G. Perfetti, *International Journal of Immunopathology and Pharmacology*, Vol. 20, p. 455, 2007.
111. A.J. Roelofs, K. Thompson, S. Gordon, and M.J. Rogers, *Clinical Cancer Research*, Vol. 12, p. 6222s, 2006.
112. Wikipedia, Mevalonic acid — wikipedia, the free encyclopedia, 2017. [Online; accessed 4-November-2017].
113. Wikipedia, Mevalonate pathway — wikipedia, the free encyclopedia, 2017. [Online; accessed 4-November-2017].
114. M. Sato and W. Grasser, *Journal of Bone and Mineral Research*, Vol. 5, p. 31, 1990.

115. H. Murakami, N. Takahashi, T. Sasaki, N. Udagawa, S. Tanaka, I. Nakamura, D. Zhang, A. Barbier, and T. Suda, *Bone*, Vol. 17, p. 137, 1995.
116. M.S. Reddy, T.W. Weatherford III, C.A. Smith, B.D. West, M.K. Jeffcoat, and T.M. Jacks, *Journal of Periodontology*, Vol. 66, p. 211, 1995.
117. S.L. Ruggiero, B. Mehrotra, T.J. Rosenberg, and S.L. Engroff, *Journal of Oral and Maxillofacial Surgery*, Vol. 62, p. 527, 2004.
118. J.V. Bagan, J. Murillo, Y. Jimenez, R. Poveda, M.A. Milian, J.M. Sanchis, F.J. Silvestre, and C. Scully, *Journal of Oral Pathology & Medicine*, Vol. 34, p. 120, 2005.
119. V. Kumar, B. Pass, S.A. Guttenberg, J. Ludlow, R.W. Emery, D.A. Tyndall, and R.J. Padilla, *The Journal of the American Dental Association*, Vol. 138, p. 602, 2007.
120. S. Cheong, S. Sun, B. Kang, O. Bezouglaia, D. Elashoff, C.E. McKenna, T.L. Aghaloo, and S. Tetradis, *Journal of Oral and Maxillofacial Surgery*, Vol. 72, p. 2461, 2014.
121. A. Yaffe, M. Iztkovich, Y. Earon, I. Alt, R. Lilov, and I. Binderman, *Journal of Periodontology*, Vol. 68, p. 884, 1997.
122. A. Yaffe, G. Golomb, E. Breuer, and I. Binderman, *Journal of Periodontology*, Vol. 71, p. 1607, 2000.
123. A. Yaffe, A. Herman, H. Bahar, and I. Binderman, *Journal of Periodontology*, Vol. 74, p. 1038, 2003.
124. T. Shibutani, A. Inuduka, I. Horiki, Y. Iwayama, and Q. Luan, *Clinical Oral Implants Research*, Vol. 12, p. 109, 2001.
125. S.A. Guelcher, J. Dumas, E.M. Prieto, A. Talley, A. Harmata, and K. Zienkiewicz, Synthetic polyurethane composite, US Patent 9 801 946, assigned to Vanderbilt University (Nashville, TN), October 31, 2017.
126. A.E. Hafeman, B. Li, T. Yoshii, K. Zienkiewicz, J.M. Davidson, and S.A. Guelcher, *Pharmaceutical Research*, Vol. 25, p. 2387, 2008.
127. M. Levy, *Bull. Israel Chem. Soc.*, Vol. 18, 1 2005.
128. W. Jia, Living polymer in situ system and method of use, US Patent 9 657 202, assigned to Gaia Dental Products, Inc. (Meriden, CT), May 23, 2017.

2

Implants

A dental implant is a root device used in dentistry to support restorations that resemble a tooth or group of teeth to replace missing teeth (1). The dental implants, abutments, and dental prostheses are collectively called dental restorations or implant systems that resemble a tooth or group of teeth, also referred to as *restoration system* or *implant system* as replacements for missing teeth.

A dental implant generally appears similar to an actual tooth root and is placed within the bone of the jaw to replace the root of the missing tooth. After the implant surface fuses with the surrounding jawbone (osseointegration), dental abutments and other dental prostheses, such as crowns, implant-supported bridges or dentures, can be installed. The dental abutments and prostheses then allow the patient to use the restorations for chewing, also called masticatory loading (1).

The process of placing the dental implants into the jawbone of a patient is called dental implantation. This is a very vigorous surgical procedure, resulting in bone damage at the bone-implant interface. A relatively long healing period follows this dental implantation process, which lasts at least about two to three months and may extend to six months. During the healing period (1):

1. The bone damage is repaired and replaced with new bone tissues (active biological bone remodeling), and
2. Direct bone ingrowth or fusion between the implant surface and the bone tissue surrounding the implant is also achieved (osseointegration).

If the healing time is too short before any masticatory force is applied on the implant, the implant might risk failure because of the bone damage in the preexisting interfacial bone, weak new bone tissues, and unstable bone-implant interface with partial osseointegration. The masticatory force applied on an insufficiently healed implant creates excessive micro-motion between bone and implant surface, resulting in fibrous tissue development at the interface, which might block further osseointegration and cause eventual failure of the implant system (1).

To prevent or to reduce any possible direct masticatory force being applied on the implant, the installed implant is protected under a healing cap during the healing period. After a sufficient healing period, a second surgery is conducted to install an abutment and prosthesis, an artificial tooth crown. The combination of these two surgeries results in an implant system that is regarded as a dental replacement for the missing tooth (1).

2.1 Dental Restoration Methods

There have been a variety of known methods devised to implant and secure a dental prosthesis. The most common type of implant is endosseous, in which an implant is first surgically placed into the patient's jawbone (2).

The implant serves to mimic a root structure and protrudes through the gum to hold an abutment adapted to receive a dental prosthesis. A common abutment is a substantially cylindrical device that is typically screwed into the implant, and the crown is then affixed on top of the abutment.

The traditional way for the dental restoration of implant includes the following steps (2):

1. Take off the sealing cap,
2. Prepare the tray, and try the impression tray on, drawing the line to cut and open the window to allow the impression coping to stick out (known as open-tray technology),
3. Put regular impression coping on, and screw tight,
4. Put impression material into the prepared tray
5. Press the impression on top of the impression coping, and wait until set,

6. Unscrew the impression coping,
7. Place wax to close the open window around the portion sticking out of the impression coping,
8. Take off the impression tray with impression coping on it,
9. Connect the implant analog with the impression coping and send all together to dental lab,
10. A laboratory technician will pour material and prepare a stone model, then send it back to the dentist,
11. The dentist chooses the abutment he or she thinks will fit the stone model best, and will send it back to the laboratory, and
12. The laboratory technician adjusts the abutment, then may send it back to the dentist to try on the patient's mouth, then readjusts the abutment again based on the dentist's prescription, and makes the crown on top of the abutment.

The common dental restoration processes are disadvantageous as they are procedurally very complicated, causing a low penetration rate for dental implants. At the moment, less than 5% of general dentists can do a surgical placement of dental implant, and less than 10% of general dentists actually are doing implanting restoration due to the complications of the current technology.

To increase the penetration rate of the implant in restoration dentistry, a dentist needs to use a simplified restoration procedure and technology, which should be straightforward, cost effective, and include a minimal number of office visits.

The need is also highlighted by the burden that in order to properly transfer an accurate model of the soft tissue in the area of the planned implant, most abutment systems now use closed tray or open tray impression technologies. These require a modification of the abutment in the lab by a dental lab technician, or the doctor uses a temporary plastic abutment, modifies it in the patient's mouth, and then sends it to a lab to cast the metal abutment base on the shaped temporary abutment (2).

There is also a monograph concerning the esthetic soft-tissue management of teeth and implants (3).

2.2 Implant Designs

Typically, a dental implant is implanted into the bone of a patient's jaw and comprises a socket, e.g., a bore, which is accessible through the overlying or surrounding gum tissue for receiving and supporting one or more attachments or components (4). These are useful to fabricate and support the prosthodontic restoration. Dental implant procedures typically involve a threaded implant that is screwed into the bone tissue.

While numerous design iterations have been marketed, overall there have been two types of implant abutment interfaces within these assemblies (4):

1. An external connection implant and an internal connection implant,
2. An internal connection implant design, and
3. A lateralized surface surrounding design.

The external hexagonal implant design typically has a hexagonal boss, or another anti-rotation feature, protruding out of the implant's upper surface, and the corresponding abutment has a female anti-rotational receptacle. The upper surface below the hexagonal boss engages the abutment. The hexagonal boss acts to constrain the abutment from rotating around the longitudinal axis as well as preventing movement on the plane coincident with the implant seating surface. Unfortunately, such an interface has very little stability until the screw is introduced and fully seated between the abutment and the implant. The screw is essentially the sole component resisting bending forces.

In contrast, the internal connection implant design has an anti-rotational female member or socket, e.g., a hexagonal anti-rotation feature, which is located below the implant's upper surface, and the corresponding abutment has a male protrusion. The abutment is typically seated on the upper surface as the external hexagonal design, the only difference being that the anti-rotation feature on the implant is located below this surface.

The benefit of this system is that it has an intrinsic stability without the screw, and then experiences increased stability once the screw is introduced and fully seated. The system responds in a more unified manner to bending forces. While this system has advantages over the external hex implant, one disadvantage, which applies to the external hex as well, is that it is prone to leak at the implant abutment interface, i.e., the seating surface, due to lifting of the abutment under load that may create an intermittent gap, resulting in bacteria penetration and a subsequent crestal bone loss.

It is known to include, at the proximal end of the implant, a lateralized surface surrounding the area where the abutment sits within the implant. In such designs, the diameter of the implant at its proximal end is larger than the diameter of the abutment in an attempt to inhibit crestal bone remodeling, i.e., the implant system is platform-switched. It is also known in the art to include a lateralized surface that is disposed at a positive slope relative to the central axis, such that the lateralized surface tapers downwards away from the uppermost region of the central bore of the implant (4).

Examples of such designs have been given. In one example (5) an outer surface tapers in a positive direction from a maximum diameter region to a smaller diameter region. In another example (6) the implant includes a step in the middle of the top surface, with the general flow being in a positive direction from a maximum diameter region to the smaller diameter step region.

In dental implant systems, small microgaps between the implant and the abutment might be present, even if the implant and the abutment are tightly sealed. Fluids may enter the small microgaps between the implant and the abutment, which is undesirable. In such systems, the positive slope of the lateralized surface tends to inhibit fluids from flowing downwards toward the central bore of the implant and into the micrograps between the implant and the abutment (4).

2.3 Dental Restoration Composition

A method for dental restoration has been presented. This method consists of the following steps (2):

1. Placing a sub-crown member comprising a final abutment component and a dental implant component into a jawbone, wherein said final abutment component comprises a modifiable portion that is made from a non-titanium material,

and the final abutment is to be customized by modifying the modifiable portion,

2. Directly measuring the shortest distance X1 between a location L1 on the surface of said modifiable portion and the surface of an object surrounding said modifiable portion in the oral environment,
3. Comparing the value X1 with a predetermined value Y1, wherein Y1>0,
4. If X1 is smaller then Y1, directly modifying, such as cutting, the modifiable portion on its location L1 in the oral environment without removing the modifiable portion out from the oral environment to increase said shortest distance X1 until X1 is equal to Y1,
5. Optionally repeating steps 2 to 4 on one or more of other locations Ln on the surface of said modifiable portion with corresponding shortest distances Xn and corresponding predetermined values Yn until Xn is greater or equal to Yn, wherein Yn>0, n is an integer, and n is greater or equal to 2, to completely customize the final abutment component,
6. Producing a crown based on the shape of the final abutment component that has been completely customized, and
7. Attaching the crown to the completely customized final abutment component in the oral environment to complete the dental restoration.

The dental implant component comprises a biostable polymeric material and a pharmaceutical agent that may be incorporated into the polymeric material.

Monomers that can be used for the polymers are collected in Table 2.1. Some of these compounds are shown in Figure 2.1.

For the fabricaton of the polymeric material, compounds such as polymerization initiators, polymerization accelerators, ultraviolet light absorbers, antioxidants, and other additives may be used.

The final polymeric material may be visible light-curable, self-curing, dual-curing, and vacuum, heat, and pressure curable compositions.

It may be fully or partially polymerized using photo, chemical or thermal means under controlled pressure or atmospheric pressure.

Table 2.1 Monomers for dental implants (2).

Monomer	Monomer
Methyl methacrylate	Ethyl methacrylate
Isopropyl methacrylate	2-Hydroxyethyl methacrylate
3-Hydroxypropyl methacrylate	2-Hydroxy-1,3-dimethacryloxy-propane
n-Butyl methacrylate	Isobutyl methacrylate
Butoxyethyl methacrylate	Hydroxypropyl methacrylate
Tetrahydrofurfuryl methacrylate	Glycidyl methacrylate
2-Methoxyethyl methacrylate	2-Ethylhexyl methacrylate
Benzyl methacrylate	Ethylene glycol dimethacrylate
Diethylene glycol dimethacrylate	Triethylene glycol dimethacrylate
Triethylene glycol trimethacrylate	Butylene glycol dimethacrylate
Neopentyl glycol dimethacrylate	1,3-Butanediol dimethacrylate
1,4-Butanediol dimethacrylate	1,6-Hexanediol dimethacrylate
Trimethylolpropane trimethacrylate	Trimethylolethane trimethacrylate
Trimethylolmethane trimethacrylate	Pentaerythritol trimethacrylate
Pentaerythritol tetramethacrylate	Poly(oxytetraethylene glycol) dimethacrylate
2,2-Bis(methacryloxyphenyl) propane	2,2-Bis(4-(2-hydroxy-3-methacryloxypropoxy)phenyl) propane
2,2-Bis(4-methacryloxydiethoxyphenyl) propane	2,2-Bis(4-methacryloxypolyethoxyphenyl) propane
Di-2-methacryloxyethyl-2,2,4-trimethylhexamethylene dicarbamate	1,3,5-Tris(1,3-bis(methacryloyloxy)-1-2-propoxycarbonylaminohexane)-1,3,5-1-(1H,3H,5H)triazin-2,4,6-trione
Urethane oligomer synthesized of 2,2′-di(4-hydroxycyclohexyl)propane, ε-caprolactone, hexamethylene diisocyanate and 2-hydroxyethyl methacrylate	Urethane oligomer synthesized of 1,3-butanediol, Hexamethylene diisocyanate and 2-Hydroxyethyl methacrylate

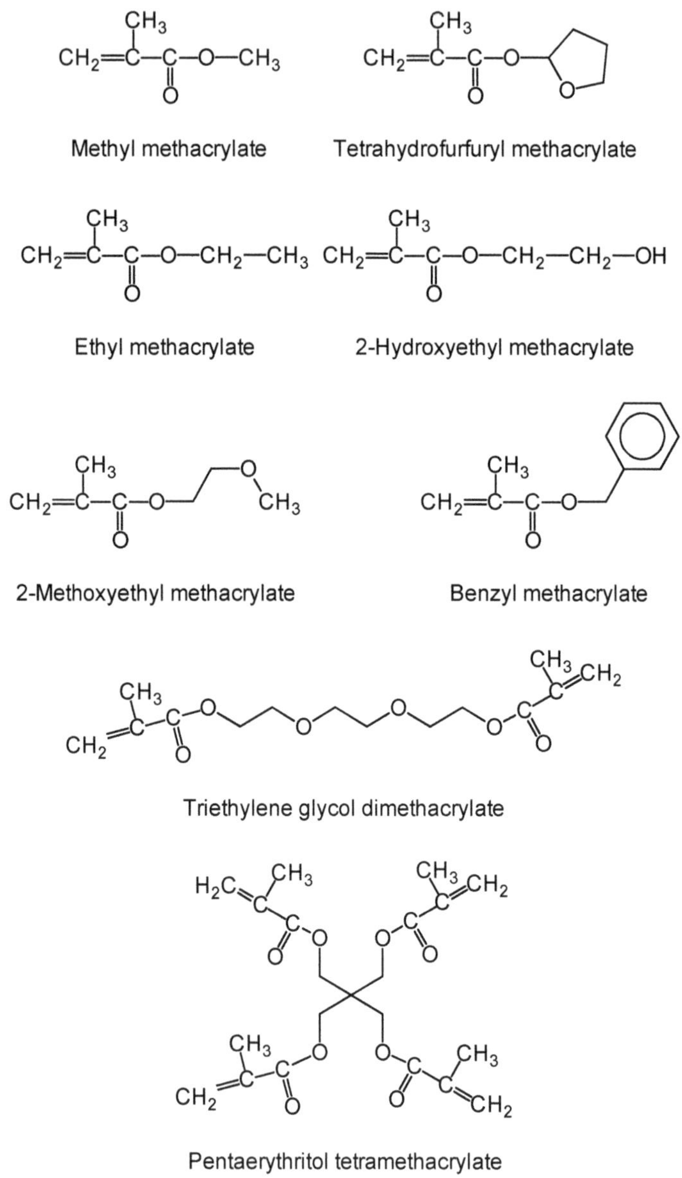

Figure 2.1 Monomers.

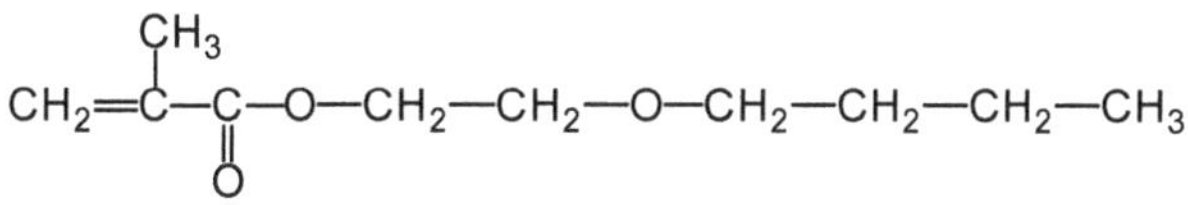

2-Butoxyethyl methacrylate

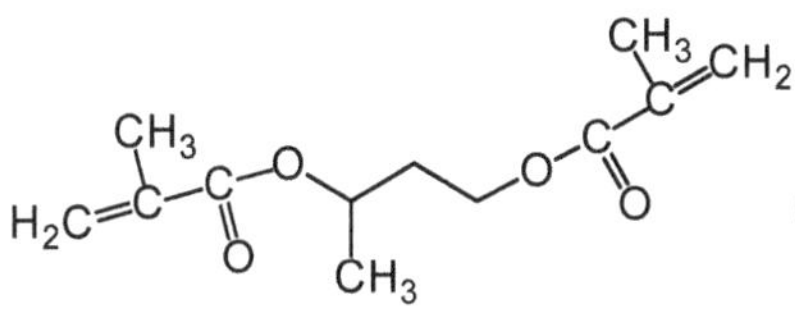

1,3-Butanediol dimethacrylate

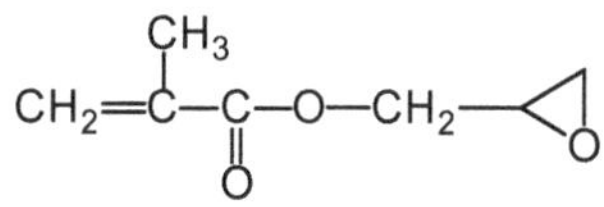

Glycidyl methacrylate

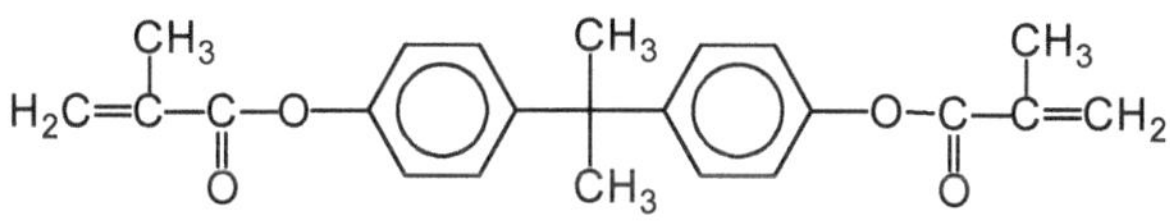

2,2-Bis(methacryloxyphenyl) propane

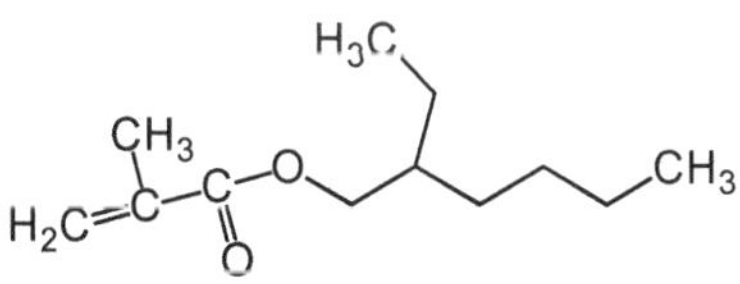

2-Ethylhexyl methacrylate

Figure 2.1 (cont.) Monomers

Visible light-curable compositions contain the usual polymerization initiators, polymerization accelerators, ultraviolet absorbers, and fluorescent whitening agents.

Preferred light-curing initiators are camphorquinone and trimethyl benzoyl phosphine oxide. Heat-curable compositions are generally filled compositions, containing a heat cure initiator, such as dibenzoyl peroxide, 1,1′-azobis(cyclohexanecarbonitrile), or other free radical initiators. These compounds are shown in Figure 2.2.

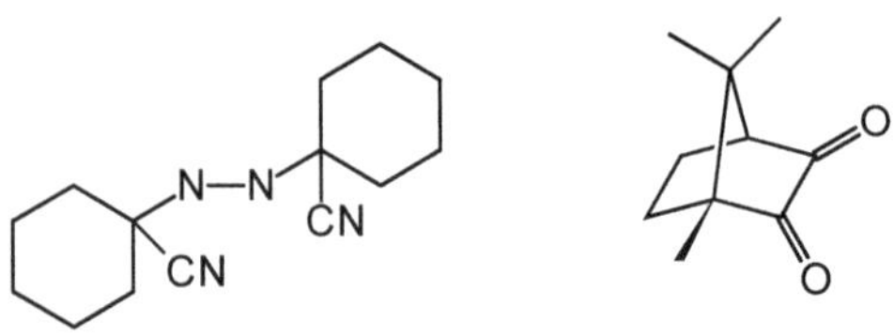

1,1′-Azobis(cyclohexanecarbonitrile) Camphorquinone

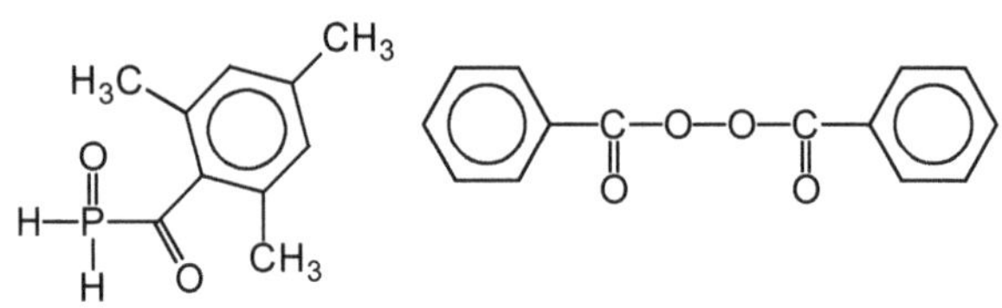

Trimethyl benzoyl phosphine oxide Dibenzoylperoxide

Figure 2.2 Curing initiators.

The pharmaceutical agent in the composition may diffuse out of the polymeric material to surrounding tissues such as the bone tissue and nearby tissue. Thus, the pharmaceutical agent may be delivered either locally or systematically by the polymeric material in the implant component. Examples of pharmaceutical agents are collected in Table 2.2 and in Table 2.3.

Antiprotozoals are shown in Figure 2.3. Antifungals are shown in Figure 2.4. Antiasthma drugs are shown in Figure 2.5. Antiviral substances are shown in Figure 2.6. Antiseptic substances are shown in Figure 2.7. Local anesthetics are shown in Figure 2.8.

Some steroidal anti-inflammatory agents are shown in Figure 2.9.

Table 2.2 Pharmaceutical agents (2).

Antiviral compounds	Antimycotic substances
Acyclovir	Cetyltrimethylammonium bromide
Antiprotozoals	Antiprotozoals
Chloramphenicol	Sulfamethoxazole
Steroidal analgesics	Anticoagulants
Aspirin Salicylic acid Diflunisal Morphine	Heparin Calcium heparin Sodium heparin Bishydroxycoumarin
Antiseptic substances Cetylpyridinium chloride Benzalkonium chloride Chlorhexidine	Vitamins Vitamin B6 Vitamin B12 Vitamin C
Antifungals	Antiasthma drugs
Poly(oxyethylene) Nonylphenols Alkylaryl sulfonates Miconazole nitrate Metronidazole Trimethoprim	Adrenaline Ephedrine Epinephrine Aminophylline Theophylline
Local anesthetics	Antihypertensive substances
Procaine Benzocaine Lidocaine Procaine Bupivacaine Tetracaine Xylocaine Mepivacaine	Methyldopa Hydralazine Clonidine Chlorothiazide Timolol Propranolol Metoprolol Prazosin hydrochloride Furosemide

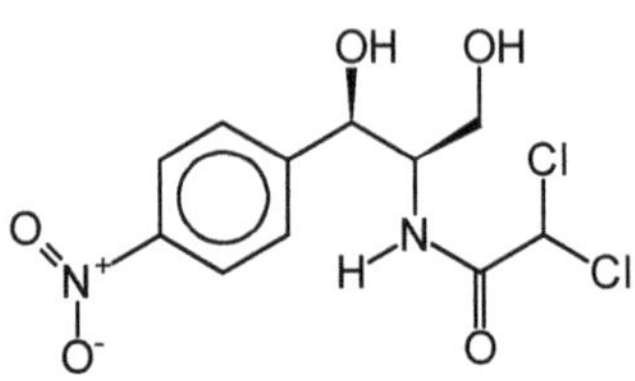

Chloramphenicol

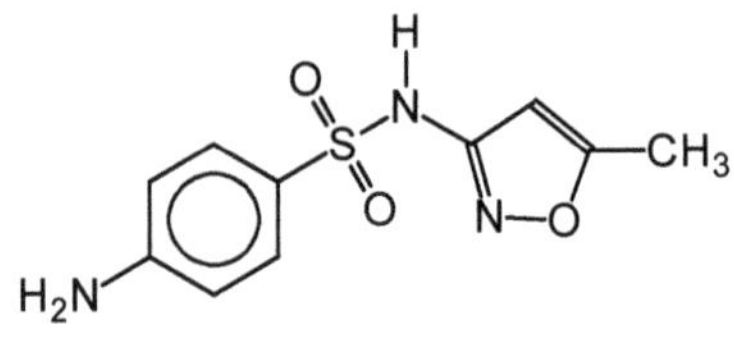

Sulfamethoxazole

Figure 2.3 Antiprotozoals.

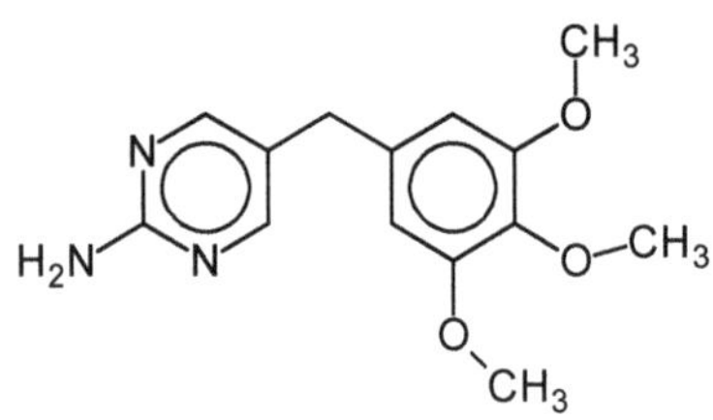

Trimethoprim

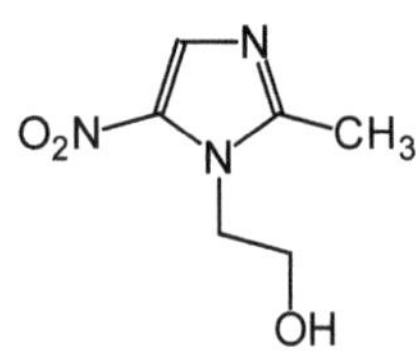

Metronidazole

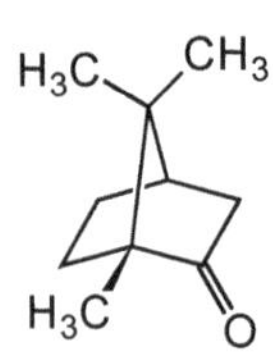

Camphor

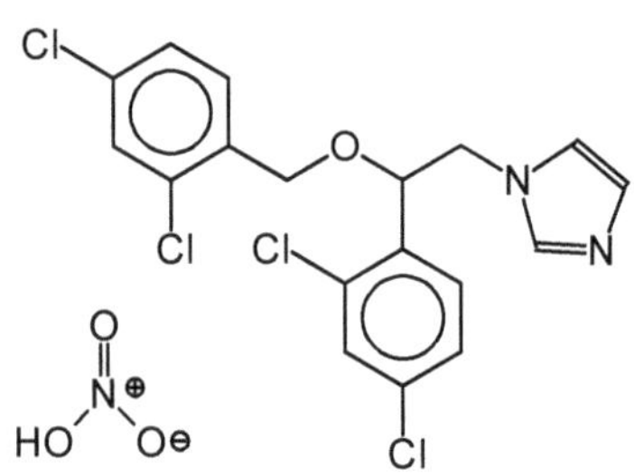

Miconazole nitrate

Figure 2.4 Antifungals.

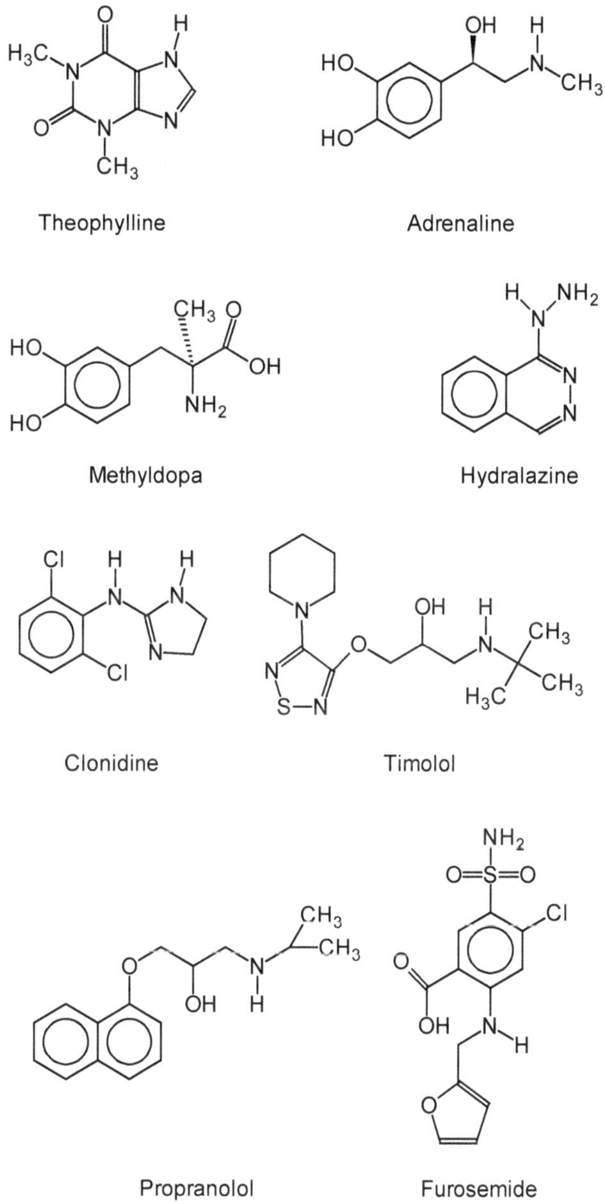

Figure 2.5 Antiasthma drugs and antihypertensive substances.

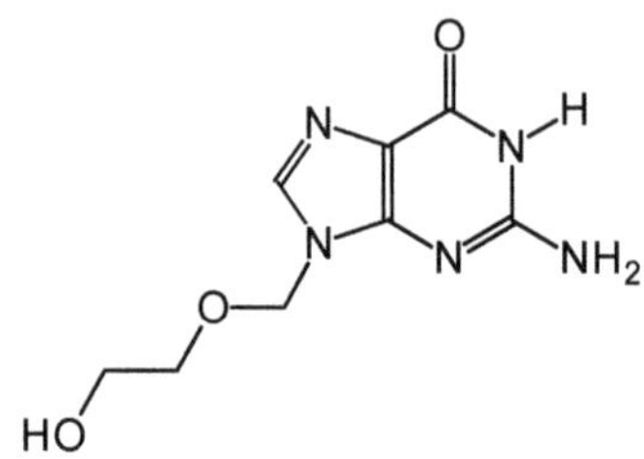

Acyclovir

Figure 2.6 Antiviral substances.

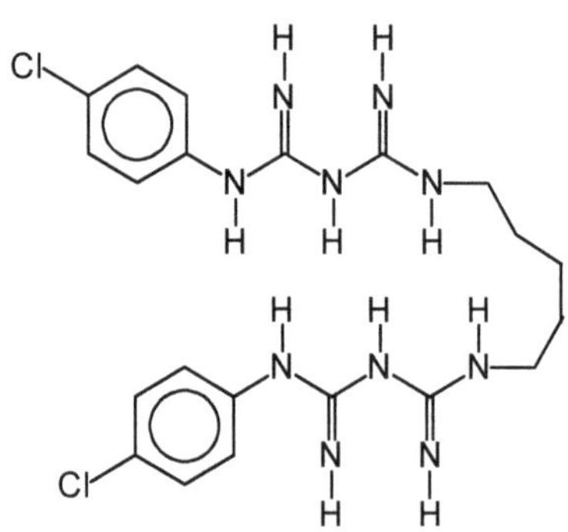

Chlorhexidine

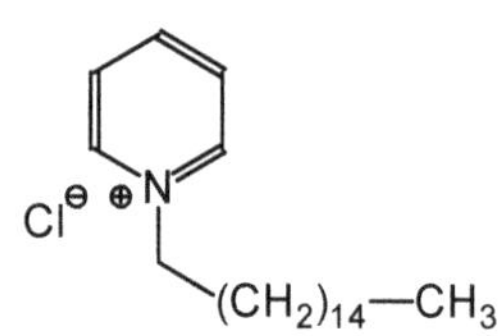

Cetylpyridinium chloride

Figure 2.7 Antiseptic substances.

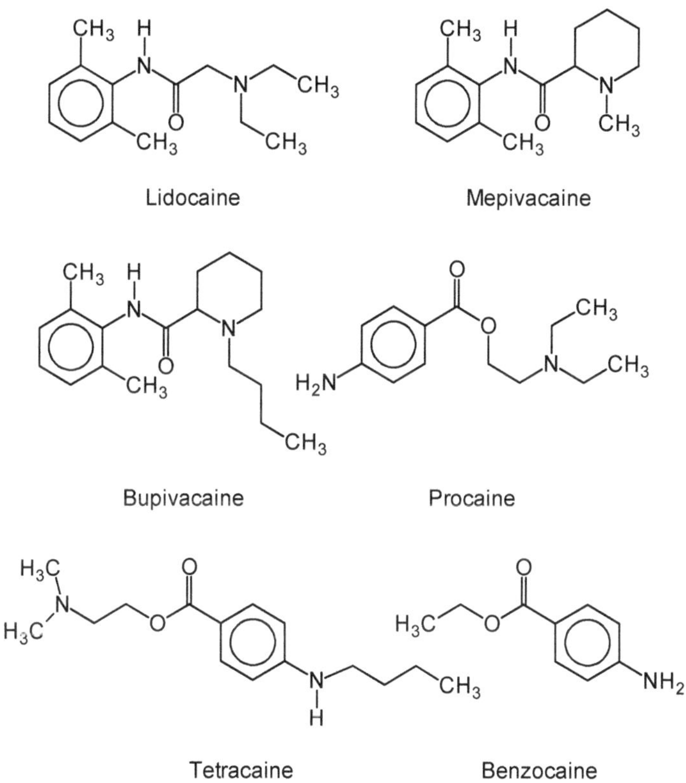

Figure 2.8 Local anesthetics.

Table 2.3 Pharmaceutical agents (2).

Steroidal anti-inflammatory agents	Steroidal anti-inflammatory agents
Hydrocortisone	Flucetonide
Hydroxyltriamcinolone	Fludrocortisone
α-methyl dexamethasone	Fluradrenolone
Dexamethasone-phosphate	Fludrocortisone
Beclomethasone dipropionates	Difluorosone diacetate
Clobetasol valerate	Fluradrenolone acetonide
Desonide	Medrysone
Desoxymethasone	Amcinafel
Desoxycorticosterone acetate	Amcinafide
Dexamethasone	Betamethasone
Dichlorisone	Chloroprednisone
Diflorasone diacetate	Chloroprednisone acetate
Diflucortolone valerate	Clocortolone
Fluadrenolone	Clescinolone
Fluclorolone acetonide	Dichlorisone
Fludrocortisone	
	Difluprednate
Flumethasone pivalate	Flucioronide
Fluosinolone acetonide	Flunisolide
Fluocinonide	Fluorometholone
Flucortine butylesters	Fluperolone
Fluocortolone	Fluprednisolone
Fluprednylidene acetate	Hydrocortisone valerate
Flurandrenolone	Hydrocortisone cyclopentylpropionate
Halcinonide	Hydrocortamate
Hydrocortisone acetate	Meprednisone
Hydrocortisone butyrate	Paramethasone
Methylprednisolone	Prednisolone
Triamcinolone acetonide	Prednisone
Cortisone	Beclomethasone dipropionate
Cortodoxone	Triamcinolone

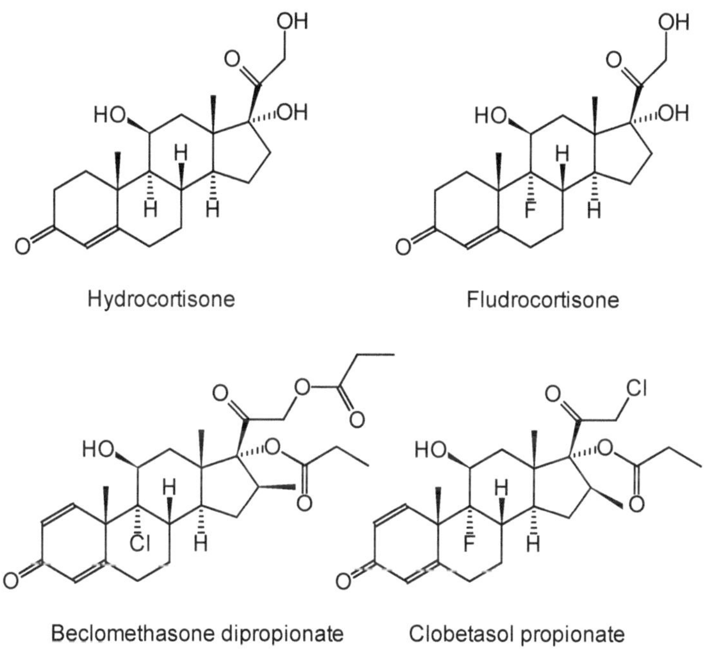

Figure 2.9 Steroidal anti-inflammatory agents.

Also, coloring agents may be formulated and added (2). A pink coloring agent may be obtained by dissolving manganese oxide in aluminum oxide as a solid solution. A yellow coloring agent may be obtained by dissolving vanadium oxide in zirconium oxide as a solid solution.

Suitable fillers are those that are capable of being covalently bonded to the polymeric matrix itself or to a coupling agent that is covalently bonded to both. Examples of suitable filling materials are shown in Table 2.4.

Table 2.4 Fillers (2).

Material	Material
Silica	Silicate glass
Quartz	Barium silicate
Strontium silicate	Barium borosilicate
Strontium borosilicate	Borosilicate
Lithium silicate	Amorphous silica
Ammoniated calcium phosphate	Deammoniated calcium phosphate
Zirconia	Tin oxide
Titania	

The reinforcing fiber element of the polymeric material preferably comprises glass, carbon, graphite, poly(aramid), or other fibers, such as polyesters, polyamides, and other natural and synthetic materials that are compatible with the polymeric matrix.

The fibers may further be treated, e.g., silanized, to enhance the bonding forces between the fibers and the polymeric matrix. The fibers preferably take the form of long, continuous filaments. Suitable coupling agents include silane compounds such as organo-silane agents. Exemplary silane agents include γ-methacryloxypropyltrimethoxysilane, γ-aminopropyl triethoxysilane, vinyl trichlorosilane and styrylamine functional silane (2). Some of these compounds are shown in Figure 2.10.

2.3.1 Heparin

A hybrid titanium composition was designed by modifying the titanium surface with heparin followed by a growth/differentiation

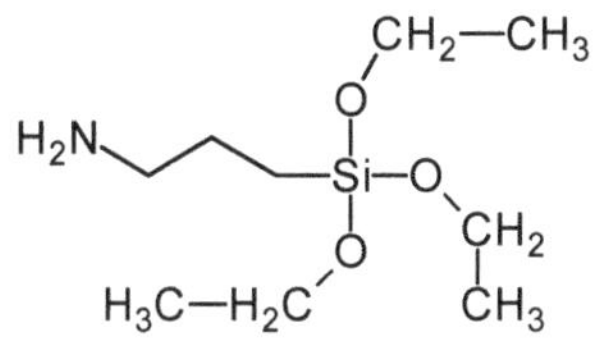

γ-Aminopropyl triethoxysilane

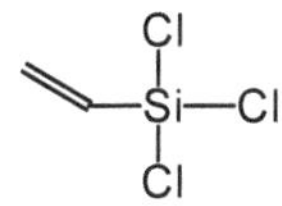

Vinyl trichlorosilane

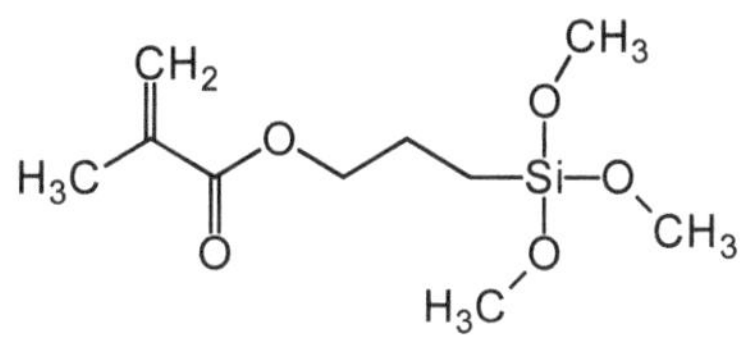

γ-Methacryloxy propyltrimethoxysilane

Figure 2.10 Silane agents.

factor-5 (GDF-5) loading (7). After that, the fabricated products were characterized by physicochemical analysis. Also, a quantitative analysis of functionalized groups was done.

The release behavior of GDF-5 grafted samples was confirmed for up to 21 *d*. The surface modification process was found to be successful and to effectively immobilize GDF-5 and provide for its sustained release behavior. As an *in-vitro* test, GDF-5 loaded Ti showed significantly enhanced osteogenic differentiation with increased calcium deposition under nontoxic conditions against periodontal ligament stem cells (7).

Furthermore, an *in-vivo* result showed that the GDF-5 loaded Ti exhibited a significant influence on new bone formation in a rabbit model. These results clearly confirmed that this strategy may suggest a useful paradigm by inducing osseointegration as a means to remodeling and healing of bone defects for restorative procedures in dentistry (7).

2.4 Hybrid Implants

A hybrid dental implant includes a screw body with one or more external threads (1). The screw body has an upper portion with an open ceiling, a middle portion having a hollow inner channel, and a lower portion with a closed floor.

The hollow inner channel is connected to the open ceiling to enable it to receive one or more injectable bone growth agents. Also, additional medical agents can be included prior to, during, or after dental implantation. Preferably, the hollow inner channel has a diameter in the range of about 1.0 *mm* to about 1.5 *mm*.

The hybrid dental contains a hydrogel loaded with bone-inducing agents. The hydrogel can be injected or introduced into the hollow inner channel in the course of the implantation. The combination of the loaded hydrogel and the dental implant provides a dental implant system that allows for controlled delivery of bone-inducing agents and other therapeutic agents to enable bone regeneration in a controlled fashion to allow for bone ingrowth into the hollow inner channel and bone outgrowth from the hollow inner channel into the surrounding bone (1).

A hydrogel is a network of polymer chains that are hydrophilic, sometimes found as a colloidal gel in which water is the dispersion medium. Hydrogels are highly absorbent, they may contain more than 90% water, and are made from natural or synthetic polymeric networks. Hydrogels also possess a degree of flexibility very similar to natural tissue, due to their significant water content. Often, it can be hardened or remain semi-liquid to a provide sustained release profile. The implant enables the use of such hydrogels without compromising the mechanical stability of the implant. As hydrogel, poly(ethylene glycol) may be used (1).

2.4.1 *Electrophoretically Prepared Hybrid Materials*

In order to get biomaterials with controllable physicochemical properties, hybrid biomaterials composed of biocompatible biopolymers and ceramic nanoparticles have been prepared (8).

The biopolymer/ceramic hybrids consisting of various natural biopolymers and layered double hydroxide ceramic nanoparticles were fabricated using an electrophoretic method. The structures and the controlled-release properties of these materials were studied.

The hybrid possessed controllable drug release properties according to the type of biopolymer included (8).

Hybrid biomaterials consisting of two or more components have been developed in order to achieve synergic effects. For example, polymer/polymer hybrids of gelatin, alginate, hyaluronate, and chitosan have been reported as wound dressings. These materials exhibit a controlled porosity and water uptake (9).

Polymer/ceramic hybrids have been prepared that consist of biopolymer hydrogels and layered ceramic nanomaterials (8).

These materials were prepared by an electrophoretic hybridization, in which ceramic particles were grown *in-situ* in a hydrogel matrix, thereby resulting in the uniform distribution of ceramic nanomaterials (10).

Several natural biopolymers, which are known to form hydrogels, and ceramic nanomaterials, i.e., a layered double hydroxide (LDH), were chosen for the fabrication.

2.4.1.1 *Layered Double Hydroxide Hybrid Materials*

Biopolymer/layered double hydroxide hybrid materials were fabricated by a electrophoresis method. Here, first, the biopolymer powder (1 wt/v% for agarose and 2 wt/v% for gelatin and the other biopolymers) was dissolved in a tris-HCl buffer with a pH of 7.4 at 120°C. Then, the solution was poured into the center of the electrophoretic kit walled by plastic plates and cooled down to room temperature for 4 *h* to obtain a cuboidal hydrogel.

The basic structure of a layered double hydroxide hybrid material is shown in Figure 2.11.

The properties of the used biopolymers are shown in Table 2.5.

Table 2.5 Properties of the hydrogel polymers (8).

Biopolymer	Hydrogel formation	Cuboidal hydrogel formation	Electrophoretic hybrid formation
Agarose	yes	yes	yes
Gelatin	yes	yes	yes
Carrageenan	yes	yes	no
Xanthan gum	yes	yes	no
Alginate	yes (with divalent cations)	no	no
Pectin	yes (with divalent cations)	no	no
Hyaluronic acid	no	no	no

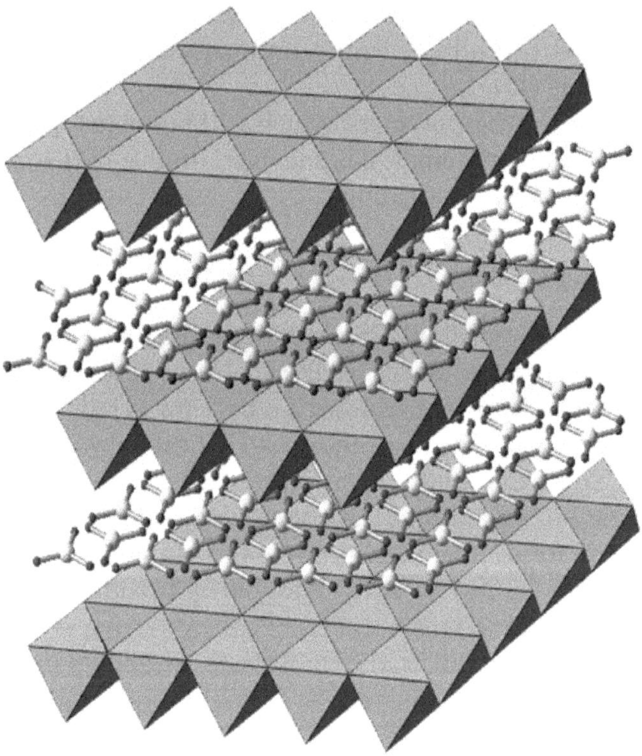

Figure 2.11 Basic structure of a layered double hydroxide hybrid material. Reprinted from (8) with permission from Springer Nature.

The cationic metal solution (0.16 *M* Zn^{2+} and 0.08 *M* Al^{3+}) and the anionic solution (0.08 *M* $NaHCO_3$ and 1 *mL* NH_4OH), which were precursors for layered double hydroxide hybrid materials, were located at each side of the cuboidal hydrogel. Then, the electrophoresis was done with 25 *V* for 30 *min*. After reaction, the hydrogel was washed with deionized water and thoroughly dehydrated (8).

The so prepared hybrid materials were characterized by X-ray diffraction (XRD), X-ray absorption spectroscopy (XAS), and field emission scanning electron microscopy (FE-SEM).

In order to identify the crystal structure of the ceramic particles in the hybrid, XRD patterns and XAS spectra were obtained by a Bruker D2 phaser with Ni-filtered Cu–K_α radiation at 1.5406 Å and at the 7D XAFS beam line at the Pohang Accelerator Laboratory (Pohang, Korea), respectively.

The FE-SEM images, which were obtained with a Hitachi SU-70 microscope at the Korea Basic Science Institute (Gangneung Center, Korea), showed the shape and size of the layered double hydroxide ceramic nanoparticles in the hybrids (8). The FE-SEM images are shown in Figure 2.12.

The various characterizations revealed that a biopolymer/layered double hydroxide hybrid with small and uniform ceramic layered double hydroxide nanoparticles can be prepared by an electrophoretic method. Because both components of the hybrid, i.e., the biopolymer and layered double hydroxide, are biomaterials, the resultant hybrid can be utilized in biomedical applications (8).

2.5 CAD/CAM Implants

2.5.1 CAD/CAM Implant-Supported Crown

A technique for immediately restoring single dental implants with a CAD/CAM implant-supported crown milled from a poly(methyl methacrylate) block has been described (11).

This technique can be used when a single dental implant is placed. A stent made of autopolymerized acrylic resin is used to transfer the implant position to the laboratory. Once the implant position has been transferred, the stone cast was scanned, and a computer-aided design and computer-aided manufacturing (CAD-CAM) interim implant-supported crown was milled from a poly(methyl

In summary, the direct application of an auto-polymerizing resin to the titanium abutment of one-piece implants increased the cervical implant surface temperature (25).

2.10 Dental Implant with Porous Body

Much effort has been directed toward integrating dental implants into the surrounding bone (26). Ideally, a dental implant would be placed into alveolar bone, and thereafter bone would readily grow into the surface of the implant.

To achieve this goal, different surface technologies have been applied to dental implants. In some instances, the surface of the implant is roughened, grit-blasted, plasma-sprayed, or microtextured. In other instances, the surface is coated with a biological agent such as hydroxylapatite (26).

2.10.1 Porous Coatings

Porous coatings have also been applied to surfaces of dental implants (27). Here, a dental implant having multiple textured surfaces on the same implant has been described. One surface includes a porous coated substrate, and another surface includes a nonporous surface adapted to encourage bone growth or bonding.

Porous coatings are advantageous since bone will indeed grow into the surface of the implant (26). Osseointegration, to a limited extent then, has been achieved with porous coated surfaces. However, these surfaces are far from ideal in terms of accepting and encouraging bone growth into the body of the implant.

Porous surfaces are often thin coatings applied to the metallic substrate of the implant. Bone surrounding the implant can only grow into the coating itself. The bone cannot grow through the coating and into the metallic substrate. The depth of bone growth into the implant is limited to the depth of the porous coating. Bone simply cannot grow completely through the implant.

A dental implant has been developed with a coronal body that connects to a bone fixation body (26). The bone fixation body has a porous structure and also includes an internal cavity with a substance that stimulates the growth of the bone. The porous structure extends through a center of the bone fixation body in a cross-sectional view of the bone fixation body.

The porous structure is formed by one or more of polymers, ceramics, and biocompatible metals and metal alloys. For example, the porous structure is constructed with tantalum, titanium, a titanium alloy, such as titanium with one or more of zirconium, niobium, tin, silicon, molybdenum, and tantalum, a biocompatible polymer, and/or a biocompatible metal or metal alloy (26).

2.10.2 *Porous Implant Material*

A dental implant is formed from: a heart material which is formed from solid-columnar titanium or titanium alloy (28). A porous layer which is arranged by the heart material. The porous layer is made by sintering a plurality of spherical grains made of titanium or titanium alloy so that a plurality of continuous holes are made between the spherical grains which are bound with each other by sintering. The spherical grains each have a surface layer of gold-titanium alloy, so that the adjacent spherical grains are bound to each other by the surface layers.

The porous implant material can be unitarily bonded to bone by easily infiltrating the bone, particularly, into the porous metal body with high porosity rate. Furthermore, since the metal body with low porosity rate is bonded, the compressive strength is large in the direction of the bonded-boundary surface. Therefore, in the bonded body, the compressive strength along the bonded-boundary surface is different from the compressive strength orthogonal to the bonded-boundary surface, so that a strength property is anisotropic as human bone. Accordingly, by implanting the porous implant material into a human body with according the anisotropic strength to a directional strength property of human bone, the stress shielding can be efficiently prevented from arising.

The porous metal bodies forming the porous implant material are produced by forming expandable slurry containing metal powder, expanding agent and the like into a sheet-shape by doctor blade method or the like, dehydrating the sheet so as to make a green sheet, and expanding the green sheet after a degreasing process and a sintering process. Furthermore, a plurality of green sheets with different mix amounts of the expanding agent from each other are made. These green sheets are layered and sintered so as to make a layered body of the porous metal bodies. Then, by pressing or rolling the layered body to compress in a layered direction, the porous implant material is produced.

As binder, methyl cellulose, hydroxypropyl methyl cellulose, hydroxyethyl methyl cellulose, carboxymethyl cellulose ammonium, ethyl cellulose, or poly(vinyl alcohol) can be used.

A plasticizer is added in order to plasticize a compact material by forming the slurry. As the plasticizer, for example, polyalcohols such as ethylene glycol, poly(ethylene glycol), glycerin; oils and fats such as sardine oil, rapeseed oil, olive oil; ethers such as petroleum ether; and esters such as diethyl phthalate, di-*n*-butyl phthalate, diethylhexyl phthalate, dioctyl phthalate, sorbitan monooleate, sorbitan trioleate, sorbitan palmitate, sorbitan stearate can be used.

As surfactants, anion surfactants such as alkyl benzene sulfonate, α-olefin sulfonate, alkyl ester sulfate, alkyl ether sulfate, alkane sulfonate; nonionic surface-active agent such as poly(ethylene glycol) derivatives, polyhydric alcohol derivatives, and ampholytic active agents can be used.

As the expanding agent, an agent which can form pores in the slurry by generating gas can be used. For example, volatile organic solvents such as pentane, neopentane, hexiane, isohexane, isoheptane, benzene, octane, toluene and the like, that is, anti-soluble hydrocarbon-system organic solvent having carbon number of 5 to 8 can be used. It is preferable that the expanding agent be contained in the expandable slurry by 0.1 to 5% by weight (28).

2.11 Implant with a Polymeric Post

Dental implants are usually designed with two pieces. The first one is the post implant and the second is the prosthetic itself (29). Post implants are designed to be fitted or screwed into the bone, as they are the anchors for the prosthetic.

The dentist will usually drill a pilot hole into the bone prior to the insertion of the implant. The implant is then fitted or screwed into the pilot hole and allowed to heal before attaching the finished prosthetic.

A clinician has several options with respect to the bone preparation. In some cases there might be insufficient bone with which to place an implant, therefore the clinician can place an artificial bone under the tissue and grow bone if necessary. Also, a clinician can choose the size and type of implant to best fit the patient's needs. If the patient has little bone into which to place an implant a clinician can choose a smaller implant.

At the end of the post is an abutment or collar, to which the prosthetic is attached. The prosthetic is usually created in a lab and usually contains a metal attachment core with a ceramic surface that by design is made to look like a tooth. The finished prosthetic is snapped or connected to the implanted post and the patient at this point has a replacement tooth that is visibly and physically existent.

The use of polymers have been elucidated, whose physical properties can be adjusted to precise specifications (29). These devices comprise both flexible and rigid structures such that unique and custom implants can be produced. The adaptability and flexibility of polymers also allow for improved implant designs where metals would be completely incompatible.

The preferred plastics to be used poly(ether ether ketone) (PEEK), Hi-lubricity nylons, impact-resistant PMMA and fluor-containing polymers. These polymers are high strength plastics that are resistant to wear and fracturing. They are also resistant to moisture and chemicals, and are biocompatible. The preferred plastic can also be selected from the group of thermoplastics that are capable of being injection molded, such that the entire implant can be injection molded.

Various polymers can also be modified in order to maximize the warranted characteristics for a dental implant. This usually means incorporating the addition of a plasticizer or filler into the plastic. Plasticizers usually impart more elasticity to the polymer, therefore rendering them more resilient. Some examples of possible plasticizers are summarized in Table 2.6 and in Figure 2.15.

Fillers can also be incorporated into the plastic materials. Fillers usually modify the wear resistance, elasticity, fracture toughness and strength of the plastic. Fillers can be composed of either powder or fiber, such as pieces of monofilament. A few examples of possible fillers would be silica, silica carbide, plastic monofilaments, carbon fiber, zirconia, alumina, borosilicate glass powder, radiopaque boro-

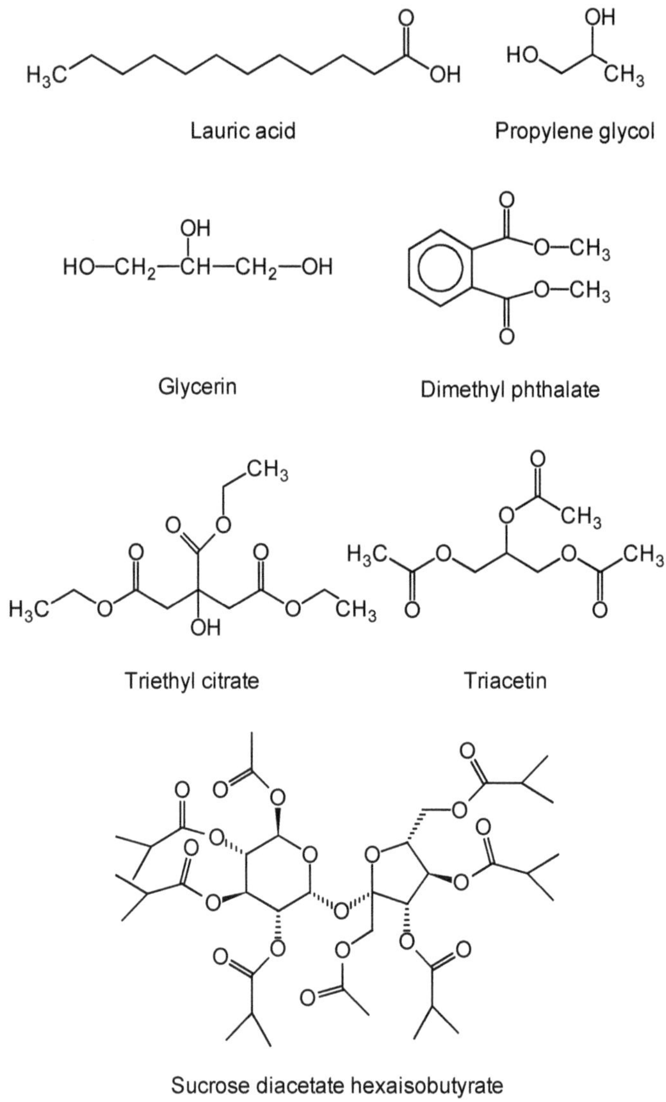

Figure 2.15 Plasticizers (29).

Table 2.6 Plasticizers (29).

Compound	Compound
Modified vegetable oils	Lauric acid
Diacetylated monoglycerides	Mineral oil
Sucrose diacetate hexaisobutyrate	Triethyl citrate
Liquid poly(ethylene glycol)s	Triacetin
Liquid poly(propylene glycol)s	Glycerin
Propylene glycol	Castor oil
Dimethyl phthalate	Dipropyl phthalate
Diethyl phthalate	Dibutyl phthalate
Dioctyl phthalate	Polysorbates
Acetyltriethyl citrate	

silicate powder, other radiopaque substances, titanium dioxide, zinc oxide, and pigments. Bioactive calcium-containing compounds may also be utilized so as to facilitate the bone growth and the bonding to the surface of the post (29).

The post implant can be designed with grooves and/or threads for an increased retention. The polymer post can have threads of any size such that the post is intended to be screwed into the bone and gains retention by biting into the bone. Another design contemplates the use of ridges or grooves that are pushed into the pilot hole and merely hold passively until bone can grow around it. The implant is shown in Figure 2.16.

A preferred polymeric post is designed with flexible retentive grooves or barbs that flex inward during insertion and resist extraction by flexing outward and biting into the bone under extraction forces. This type of post design would be impossible for a rigid metal. As such metals cannot flex, the corresponding polymeric designs represent an advantage over prior art designs.

The posts can be easily inserted and do not require a wrench or other tool to screw them in, they are simply set into place with sufficient force. Such a design is superior to a screw type design, a screw type design must bite into the hard and soft tissue where the leading edge must cut into and displace room for an enlarging thread. This cuts and destroys the bone in the process and the unattached excess tissue gets pushed between the threads. The body must remove this excess tissue before healing can begin. The

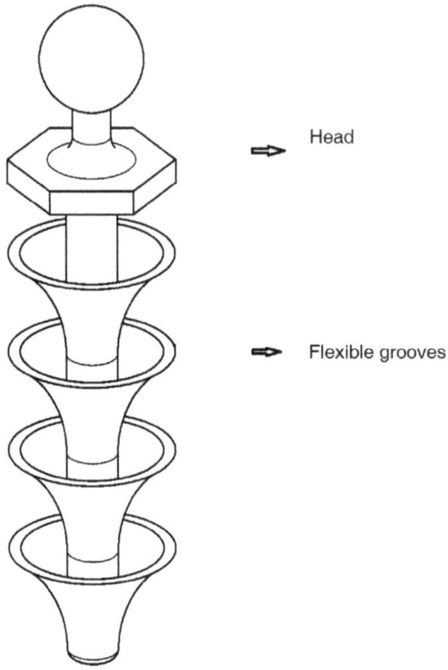

Figure 2.16 Post implant (29).

flexible groove retentive post developed here avoids the cutting and displacement of the screw and minimizes the damage done to the tissues (29).

2.12 Short Dental Implants Versus Standard Dental Implants

A systematic review and meta-analysis was done in order to compare short implants of equal or less than 8 *mm* versus standard implants larger than 8 *mm*. The implants are placed in posterior regions of maxilla and mandible. The survival rates of implants were evaluated, as well as marginal bone loss, complications and prosthesis failures (30).

The results of the literature review showed that there was no significant difference in the properties, which are shown in Table 2.7.

Short implants are considered a predictable treatment for poste-

Table 2.7 Statistical properties (30).

Item	Probability	Relative risk	Confidence interval
Implants survival	0.24	1.35	0.82–2.22
Marginal bone loss	0.06	-0.20	-0.41–0.00
Complications	0.08	0.54	0.27–1.09
Prosthesis failures	0.92	0.96	0.44–2.09

rior jaws. However, short implants with length less than 8 *mm*, ie., 4–7 *mm*, should be used with caution because they present greater risks to failures in comparison to standard implants (30).

2.13 Adjustable Dental Implants

Osseointegration dominates the bonding strength between the alveolar bone and a dental implant, whether a one-piece dental implant or a two-piece dental implant (31). Take the two-piece dental implant for instance. The dental implant comprises a fixture and an abutment. The fixture is screwed to the alveolar bone. The bottom of the abutment is screwed to the top of the fixture. The top of the abutment extends beyond the gum flap that covers the alveolar bone, so as to mount a dental crown.

As the dental implant is integrated with the alveolar bone only by the fixture in a dental implantation surgery, the quality and quantity of the alveolar bone has a great influence on the success and safety of the dental implantation surgery. Clinically, an alveolar bone at least 10 *mm* in height and at least 5 *mm* in width leads to a successful and safe dental implantation surgery. However, the tooth is extracted or exfoliates before the dental implantation surgery. The alveolar bone atrophies or is lost, or the sinus lifts after the tooth extraction and the tooth exfoliation, such that the quality and quantity of the alveolar bone declines, introducing the following issues (31):

1. The dental implant is likely to damage the tissues near the alveolar bone, resulting in the failure of the dental implantation surgery.
2. The dental implant cannot be firmly integrated with the alveolar bone and the osseointegration is not ideal. The dental implant and the alveolar bone are loosened after a period of

applying the occlusal force on the dental crown, causing the falling out of the dental implant and the dental crown and the failure of the dental implantation.

A dental implant has been developed composed of a fixture, an abutment set at the top of the fixture, and a connecting structure set between the fixture and the abutment. The connecting structure comprises two opposite wings. Each of the wings is at a predetermined angle toward the fixture. The fixture is implanted into the alveolar bone for osseointegration. The top of the abutment extends beyond the gum flap for mounting a dental crown thereon. The wings cover the surface of the cortical bone of the alveolar bone and are fixed with the cortical bone by dental posts (31).

2.14 Materials for Implants

2.14.1 Poly(ether ether ketone)

Poly(ether ether ketone) (PEEK) is a polymer that has many potential uses in dentistry. The recent research conducted on the material for dental applications has been reviewed (32). The research papers in the last fifteen years were considered.

Also, the use of PEEK as a dental implant abutment has been reviewed (33). Clinical reports have been presented to suggest some uses of PEEK materials in implant dentistry.

PEEK has been investigated for a number of applications for clinical dentistry. For example, PEEK dental implants have exhibited lesser stress shielding in comparison to titanium dental implants due to a closer match of the mechanical properties of PEEK and bone. Thus, PEEK is a promising material for a number of removable and fixed prosthesis. Furthermore, recent studies have focused improving the bioactivity of PEEK implants in the nanoscale range (32).

A common synthesis route for PEEK is the reaction between 4,4′-difluorobenzophenone and the disodium salt of hydroquinone in a polar solvent, e.g., diphenyl sulfone at 300°C. The structure of PEEK is shown in Figure 2.17.

The lower Young's elastic modulus of 3–4 *GPa* of PEEK, which is close to human bone, is beneficial for orthopedic implant applications. PEEK can be easily modified by the incorporation of other

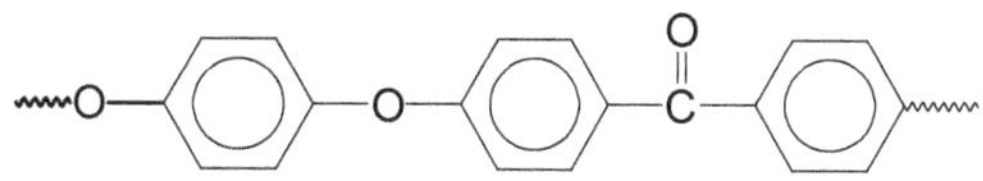

Figure 2.17 Poly(ether ether ketone).

materials. The incorporation of carbon fibers can increase the elastic modulus up to 18 *GPa*. Thus, the modulus of carbon-reinforced PEEK is comparable to that of cortical bone and dentin.

The mechanical properties of various PEEK compounds have been evaluated using the three-point bending test (34). In total, 150 specimens of 11 different PEEK compounds were tested using a three-point bending test by a universal testing machine (Z010, Zwick GmbH & Co, Ulm, Germany) after dry storage on the one hand and after incubation at 37°C in Ringer solution (Fa. Braun Melsungen AG, Melsungen, Deutschland) for 1 *d*, 7 *d*, 28 *d* and 84 *d*.

The results regarding the bending moduli ranged from 2.73 ± 0.26 *GPa* for an unfilled brand to 47.27 ± 10.3 *GPa* for a carbon fiber reinforced PEEK with unidirectional continuous carbon fibers. Accordingly, the bending strengths of these two material types ranged from 170.37 ± 19.31 *MPa* to 1009.63 ± 107.33 *MPa* (34).

All the tested specimens showed higher values than the prevailing minimum strength for plastic materials and their application in dentistry of 65 *MPa* (35). This underlines the applicability of PEEK in dentistry and points out the possibility to offer patients metal-free restorations, especially in the presence of allergies and/or bruxism (34).

The tensile strength and the elastic moduli of PEEK, other polymers and mineralized human tissues are shown in Table 2.8.

Unmodified PEEK is hydrophobic and has a water-contact angle of 80°–90° and is bioinert.

Recent proteomic studies indicate that PEEK inhibits mRNA processing (36). This may result in a decreased cellular proliferation rate on the surface and cytotoxic effects may be produced.

On the other hand, a few studies suggested that there is no significant difference between the osseointegration of PEEK and conventional implant materials such as zirconia and titanium (37,38).

Table 2.8 Tensile strength and the elastic moduli (32).

Material	Tensile strength /[*MPa*]	Young's modulus /[*GPa*]
PEEK	80	3-4
Carbon-fiber reinforced PEEK	120	18
Cortical bone	104-121	14
PMMA	48-76	3-5
Dentin	104	15
Enamel	47.5	40-83
Titanium	954-976	102-110

2.14.1.1 Bonding to Human Dentin

The effect of different surface treatments on the bond strength of PEEK bonded to human dentin was evaluated (39). One hundred PEEK cylinders (3 *mm* × 3 *mm*) were divided into five groups according to the surface treatment:

1. Silica coating,
2. Sandblasting with 45 μm Al_2O_3 particles,
3. Etching with 98% sulfuric acid for 5 *s*,
4. Etching with 98% sulfuric acid for 30 *s*, and
5. Etching with 98% sulfuric acid for 60 *s*.

These cylinders were luted with resin cement onto 50 human molars. First, each tooth was embedded in epoxy resin and the buccal dentin surface was exposed.

Then, two delimited dentin areas (3 *mm*) per tooth were etched with 35% phosphoric acid and bonded with a two-step self-priming adhesive system. After the luting procedure the specimens were stored in water for 24 *h* at 37°C. The shear bond strength was tested using a universal testing machine and the failure types were assessed.

The stress data were analyzed using the Kruskal-Wallis test. A comparison of the proportions of different failure types was performed using the Bonferroni method (p less than 0.05).

The Kruskal-Wallis test demonstrated that differences among groups were not significant ($p=0.187$). The mean shear bond strengths were found as follows: silica coating, 2.12±1.12 *MPa*; sandblasting, 2.37±0.86 *MPa*; sulfuric acid 5 *s*, 2.28±1.75 *MPa*; sulfuric acid 30 *s*, 1.80±0.85 *MPa*; sulfuric acid 60 *s*, 1.67±0.94 *MPa*.

Adhesive and mixed failures were predominant in all the tested groups. Both physical and chemical surface treatments produced adhesion between PEEK, resin cement and dentin (39).

2.14.1.2 Pressure Behavior

The pressure behavior of different PEEK materials for dental implants was tested (40).

Static pressure tests were conducted with 11 PEEK materials: two unfilled grades, two grades filled with titanium dioxide powder, two grades filled with barium sulfate powder, two grades reinforced with short carbon fibers, one grade reinforced with glass fibers and two grades reinforced with continuous carbon fibers.

The samples were in the form of cylindrical specimens with a diameter of 4, 5 and 6 *mm*. The specimens had a height to diameter ratio. The elastic moduli ranged between 2.65±0.03 *GPa* for specimens of a titanium-dioxide-filled grade and 106.71±14.83 *GPa* for specimens reinforced with continuous carbon fibers. The lowest pressure strength of 122.77 *MPa* was observed for specimens of an unfilled grade. On the other hand, the highest pressure strength of 712.67±66.02 *MPa* could be evaluated for specimens containing continuous carbon fibers (40).

2.14.1.3 Fatigue Limits

A major requirement for dental implant materials is their resistance to cyclic loading due to mastication (41). A special fatigue test was designed to evaluate the elastic behavior and long-term form stability of cylindrically shaped PEEK specimens of 4, 5 and 6 *mm* in diameter. Eleven different PEEK materials of various grades were used: unfilled, filled with titanium dioxide or barium sulfate powder, reinforced with short carbon fibers or short glass fibers, and reinforced with continuous carbon fibers.

The samples were exposed to cyclic loads of up to 2000 *N* (41). The elastic limits ranged between 46.57±6.44 *MPa* for short carbon fiber reinforced samples of 6 *mm* diameter and 107.62±8.23 *MPa* for samples of a different short carbon fiber reinforced PEEK compound with a diameter of 4 *mm*. The elastic limits of the two PEEK grades containing continuous carbon fibers could not be tested because they exceeded the limits of the test specification.

The elastic moduli ranged between 2.06±0.18 *GPa* for barium sulfate powder-filled PEEK-specimens of 6 *mm* diameter and 57.53±14.3 *GPa* for continuous carbon fibers reinforced PEEK specimens with a diameter of 4 *mm*. With regard to the elastic limit, all the investigated PEEK materials were able to resist the pressure caused by maximum masticatory forces (41).

2.14.1.4 Micro-scale Abrasion Wear

In the oral cavity, the abrasive wear is predictable at exposed tooth or restorative surfaces during mastication and tooth brushing (42). In addition, wear can occur at the contacting surfaces between the titanium-based prosthetic structures and implants in the presence of abrasive compounds from food or toothpaste.

The abrasive wear resistance of PEEK and Ti_6Al_4V on three-body abrasion with different hydrated silica contents and loads was investigated. The PEEK samples were synthesized by hot-pressing technique.

Surfaces of Ti_6Al_4V or PEEK cylinders with a diameter of 8 *mm* diameter and a height of 4 *mm* were wet ground on SiC papers and then polished with 1 μm diamond paste. Afterwards, the surfaces were ultrasonically cleaned in propyl alcohol for 15 *min* and then in distilled water for 10 *min*. Micro-scale abrasion tests were performed at 60 *rpm* and on different normal loads (0.4, 0.8 or 1.2 *N*) after 600 ball revolutions using suspensions with different weight contents of hydrated silica (42).

After the abrasive tests, wear scars on flat samples were measured to quantify the wear volume and characterized by SEM to identify the dominant wear mechanisms.

The results showed a higher volume loss rate on PEEK than that observed on Ti_6Al_4V when subjected to three-body abrasion tests involving hydrated silica suspensions. An increase in volume loss was noted on both tested materials when the abrasive content or load was increased. PEEK showed less wear resistance than Ti_6Al_4V after micro-scale abrasion wear in contact with hydrated silica particles, as commonly found in toothpastes (42).

In order to improve the simultaneous wear and corrosion response of highly biocompatible prosthetic structures, a veneering PEEK to Ti_6Al_4V substrate was assessed by tribocorrosion analysis under conditions that mimic the oral environment (43).

Tribocorrosion tests on Ti_6Al_4V or PEEK/Ti_6Al_4V samples were performed on a reciprocating ball-on-plate tribometer at 30 N normal load, 1 Hz and stroke length of 3 mm. The tests were carried out in artificial saliva at 37°C. The open circuit potential was measured before, during and after reciprocating sliding tests. The worn surfaces were characterized by SEM. The results showed a lower wear rate on PEEK combined with a lower coefficient of friction in comparison to Ti_6Al_4V. Actually, PEEK protects a Ti_6Al_4V substrate against the corrosive environment and wear, thus avoiding the release of metallic ions to the surrounding environment (43).

2.14.1.5 *Interim Fixed Dental Prosthesis*

A PEEK framework veneered with composite resin can be used as an alternative material for the fabrication of an interim 3-pontic resin-bonded fixed dental prosthesis after implant placement (44). The low modulus of elasticity, i.e., 4 GPa, of PEEK combined with the use of the indirect-light-polymerized resin as a veneering material can be used for a resin-bonded fixed dental prosthesis.

This provides an advantage over metal ceramics or ceramics in dampening the occlusal forces and reducing debonding rates (44).

2.14.1.6 *Abutment Screws*

In cases of screw-retained implant-supported reconstructions of PEEK, an abutment screw made of PEEK might be advantageous over a conventional metal screw due to its similar elasticity (45). Also, in case of abutment screw fracture, a screw of PEEK can be more easily removed.

M1.6-abutment screws with four different PEEK containing compositions were investigated using tensile tests to set their maximum tensile strengths in relation to an equivalent stress of 186 MPa, which is caused by a tightening torque of 15 $N\,cm$.

Two screw types were manufactured via injection molding and contained 15% short carbon fibers (sCF-15) and 40% (sCF-40), respectively. Two other screw types were manufactured via milling and which contained 20% TiO_2 powder (TiO_2-20) and more than 50% parallel orientated, continuous carbon fibers (cCF-50). A conventional abutment screw of Ti_6Al_4V, i.e., CAMLOG® abutment screw, CAMLOG, Wimsheim, Germany (46), was used as control.

The maximum tensile strength was 76.08±5.50 *MPa* for TiO_2-20, 152.67±15.83 *MPa* for sCF-15, 157.29±20.11 *MPa* for sCF-40 and 191.69±36.33 *MPa* for cCF-50. The maximum tensile strength of the Ti screws amounted to 1196.29±21.4 *MPa*. The results of the TiO_2-20 and the titanium screws were significantly different from the results of the other samples, respectively. For the manufacturing of PEEK abutment screws, PEEK reinforced by more than 50% continuous carbon fibers would be the material of choice (45).

2.14.1.7 *Nano-level Modification*

In order to improve its bioactivity and osseoconductive properties, PEEK has been modified at the nano-level. In detail, PEEK has been coated by bioactive materials such as osseoconductive calcium hydroxyapatite or titanium by means of plasma spraying (47). The biological response of carbon fiber reinforced PEEK implants, with and without hydroxyapatite coating, was investigated.

The results of this study demonstrated that an hydroxyapatite coating significantly enhances the bone response to PEEK implants *in-vivo*. Moreover, in a cortical bone, hydroxyapatite-coated PEEK implants induced superior bone response compared with hydroxyapatite-coated titanium ones. These results suggested that hydroxyapatite-coated carbon fiber reinforced PEEK is a suitable material for *in-vivo* implantation (47).

2.14.1.8 *Spin Coating*

Spin coating is another method for coating a thin layer of nanoscale calcium hydroxyapatite on a PEEK surface (48,49).

Here, apatite that is dissolved in organic solvents is slowly dropped onto the surface of an implant that is rotating at a high speed (48). In the course of a subsequent heat treatment a thin layer of hydroxyapatite is formed on the implants.

In a study, 48 threaded, non-cutting PEEK implants were inserted bilaterally in the tibia of 24 rabbits. Half of the implants were spin-coated with nanocrystalline hydroxyapatite and the remaining implants were left uncoated. Half of the animals were euthanized after 3 weeks of healing and the remaining animals were euthanized after 12 weeks. The implant retention was measured with a removal torque apparatus. Surface analysis was performed with interferometry, scanning electron microscopy, and X-ray photon spectroscopy to relate the removal torque to the applied surface. The test implants revealed a significantly higher retention both after 3 weeks and 12 weeks. Thus, the addition of nanocrystalline hydroxyapatite coating to PEEK surfaces significantly increases its removal torque and biocompatibility (50).

In summary, spin-coated PEEK implants show a higher bone-implant contact in comparison to an uncoated PEEK material. On the other hand, no significant differences in the removal have been observed (48, 50).

2.14.1.9 *Titanium Plasma Spray Coating*

The coating of PEEK with rough and porous titanium plasma spray coatings is a technique which is commonly used to enhance the osseointegrative properties of medical implants (51).

However, the influence of the titanium plasma spray coating on the PEEK mechanical properties has not been sufficiently tested in the past. Therefore, PEEK samples were coated with a thick titanium plasma spray layer with grains of 90 μm and 180 μm diameter.

The coating characteristics and the adhesive strength of the coatings on the samples were elucidated and compared with coatings on titanium samples. The influence of the coating process on the mechanical and chemical-physical properties of PEEK was also evaluated. All the titanium plasma spray coatings on PEEK and titanium fulfilled the requirements for thickness (200 ± 50 μm), porosity (30 ± 10%) and roughness (90 μm grain diameter coating, i.e., 25 ± 5 μm and 180 μm grain diameter coating: 45 ± 15 μm) and were able to meet the demands required for adhesive strength (greater than 22 *MPa*) and had a shear strength greater than 20 *MPa*.

It was found that the mechanical properties, such as yield stress, fracture strain, flexural modulus and flexural stress, of the PEEK samples were influenced by the coating process, while the chemical-physical properties were not altered (51).

2.14.1.10 *Plasma-Gas Etching*

Nanoscale surface modifications of PEEK also can be produced by a plasma-gas etching method (52). In a study, oxygen and ammonia plasma treatments were used. These treatments resulted in surface modifications, leading to changes in the nanostructure, contact angle, electrochemical properties, and protein adhesion.

In order to evaluate the effect of the plasma-induced PEEK modifications on stem cell adhesion and differentiation, adipose tissue-derived mesenchymal stem cells were seeded on the PEEK samples. An increased adhesion, proliferation, and osteogenic differentiation were found (52).

When the adipose tissue-derived mesenchymal stem cells were grown on 10 *W* and 50 *W* oxygen and ammonia plasma-treated PEEK substrates, they exhibited double the degree of mineralization in comparison to an original PEEK sample. Thus, the plasma treatment of PEEK specimens induce changes in the surface chemistry and topography and can support an *in-vitro* osteogenic differentiation of adipose tissue-derived mesenchymal stem cells (52).

An anodized nanoporous layer of e-beam coated titanium can carry immobilized bone morphogenic protein-2 (53). Bone morphogenic protein-2 is a growth factor that has been shown to increase the bone deposition in implant sites.

2.14.1.11 *Sulfuric Acid Etching*

Nanostructured PEEK surfaces that have been produced by etching with sulfuric acid and rinsing afterwards with distilled water have been shown to induce an accelerated osseointegration in comparison to an unmodified PEEK both *in-vitro* and *in-vivo* (54).

The results of this study revealed that the pre-osteoblast functions, bone growth, and apatite formation on the sulfonated PEEK surfaces are affected by many factors, including positive effects introduced by the 3D porous structure and $-SO_3H$ groups, as well as negative ones due to the low pH environment. So, the surface functionalization broadens the use of PEEK in orthopedic implants (54).

On the other hand, it has also been observed that there is a decreased initial bone formation when sulfonated PEEK is implanted *in-vivo* (54). This could be explained by the presence of residual sulfuric acid on the surface. However, this behavior can be minimized by rinsing the implants with acetone to remove the residual sulfuric acid.

2.14.1.12 Alkaline Treatment

Strong alkali materials can be used to introduce hydroxyl functional groups (–OH groups) on a PEEK surface. These groups help to induce the formation of a biomimetic apatite layer when immersed in a simulated biomimetic fluid (55). The apatite growth has been found to be dependent on both the concentration of the NaOH solution and the structure of the polymer surface.

Conventional biomimetic coatings can take up to 48 *h* to be produced. A microwave-assisted biomimetic coating can produce an apatite layer on PEEK within 4 *min* (56).

2.14.1.13 Melt Blending and Compression-Molding

To increase the bioactivity, bioactive inorganic particles have been incorporated into PEEK by melt-blending and compression-molding techniques (57). The amount of hydroxyapatite incorporated into the PEEK polymer matrix was 5–40 vol%. These materials could be successfully fabricated by injection molding.

The incorporation of nano-sized particles of hydroxyfluorapatite has been suggested to impart the antimicrobial properties against *Streptococcus mutans*, a common oral pathogen (58).

The amount of hydroxyapatite in the composite influenced the tensile properties. The dynamic behavior under tension-tension fatigue revealed that the fatigue life of the composites was dependent on the hydroxyapatite content as well as the applied load. The biological responses of the PEEK hydroxyapatite composites that were carried out *in-vivo* verified the biocompatibility and bioactive nature of the composite materials (57).

2.14.1.14 *Ternary Biocomposite with Enhanced Antibacterial Activity*

A carbon fiber reinforced PEEK possesses biomechanical properties, such as elastic modulus, that are similar to human bones and therefore is becoming a dominant alternative to replace the traditional metallic implants. The defective osseointegration and bacterial infection risk of carbon fiber reinforced PEEK, however, impede its clinical adoption (59).

A newly developed carbon fiber reinforced PEEK nanohydroxyapatite ternary biocomposite was functionalized by covalent grafting of carboxymethyl chitosan, cf. Figure 2.18, followed by the decoration of a bone-forming peptide assisted via the poly(dopamine) tag strategy (59).

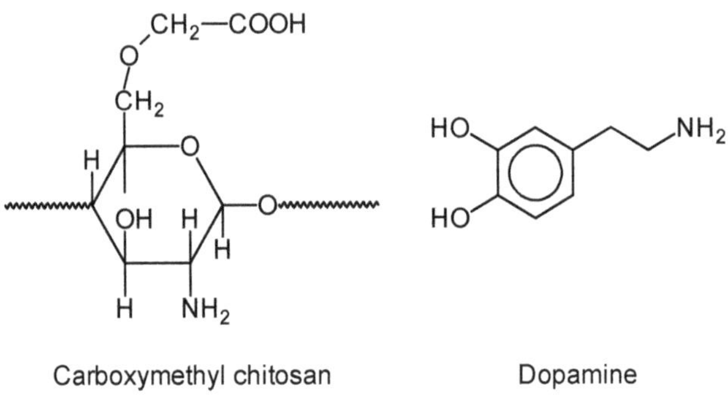

Figure 2.18 Compounds for biocomposite (59).

An antibacterial test with *Staphylococcus aureus* indicated that the carboxymethyl chitosan and the peptide-conjugated substrates significantly suppressed the bacterial adhesion. *In-vitro* cell attachment/growth, spreading assay, alkaline phosphatase activity, real-time polymerase chain reaction analysis, osteogenesis-related protein expression and calcium mineral deposition disclosed a greatly accelerated adhesion, proliferation and osteo-differentiation of human mesenchymal stem cells on the biocomposite due to the additive effect of the carboxymethyl chitosan polysaccharide and the small osteoinductive peptide. *In-vivo* evaluation of the beagle tibia model by means of micro-computed tomography, histological analysis, SEM observation and fluorescent labeling confirmed a remarkably boosted bioactivity and osteointegration (59).

Thus, the ternary composite with the dual functions of bacterial adhesion reduction and osteointegration promotion has great potential as a bioactive implant material in orthopedic and dental applications (59).

2.14.2 *Expandable Polymer Dental Implant*

A dental implant system that is suitable for an endosteal implant into a jawbone has been presented (60). The expandable polymer dental implant is shown in Figure 2.19.

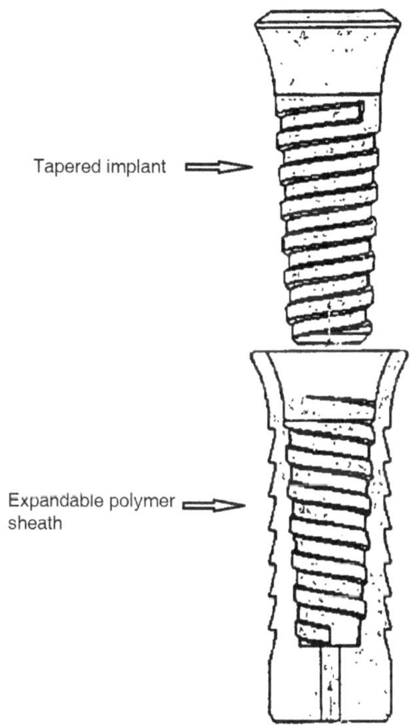

Figure 2.19 Expandable polymer dental implant (60).

The polymer is selected from ultra-high molecular weight poly(ethylene), poly(propylene), or poly(urethane) elastomers. The system consists of a tapered expandable polymer sheath, a tapered implant, and an abutment that permits the attachment of a dental prosthesis. These components are designed and constructed to form a system that provides for dynamic response and immediate load-bearing of the implant after its installation. The design of the interior of the sheath is complementary to the exterior of the implant.

The sheath serves as an artificial periodontal membrane. The second component is a rigid implant that is inserted within the polymer sheath and causes expansion of the polymer sheath when fitted within the sheath. An abutment is provided to couple the rigid implant and permits attachment of a dental prosthesis. Both components are tapered to provide for lateral transfer of loading forces onto the surrounding bone. The dental prosthesis is adapted to be used for a single tooth or to extend as a bridge over a gap (60).

2.14.3 *Endosseous Dental Implant Assembly*

A dental implant assembly has been described that contains a bio-supportive or biodegradable scaffold carried by a dental implant and impregnated with regenerative stem cells, autogenous cells, growth factors or bone graft material (61).

The growth-promoting regenerative product contains any mixed type of periodontal ligament mesenchymal stem cells which are collected in Table 2.9.

Table 2.9 Stem Cells (61).

Cell Type
Periodontal ligament mesenchymal stem cells
Dental pulp stem cells
Stem cells from exfoliated deciduous teeth
Gingival mesenchymal stem cells
Periodontal ligament stem cells
Dental follicle progenitor cells
Endothelial stem cells
Autogenous treated dentin matrix
Platelet-rich fibrin
Stem cells from apical papilla

These materials could support cementum/periodontal ligament, dentin or periodontal ligament-like tissue regeneration with neovascularization or including any osteoconductive ceramic bone powders which would support new bone osseointegration (61).

The dental implant consists of an abutment portion for connecting to a tooth crown and a hollow base portion defining a cavity therein. The abutment portion is integral with or non-movably secured to the hollow base portion. A bio-supportive or biodegradable scaffold is carried by the hollow base portion. The scaffold consists of a hollow annular body comprising an annular outer shell, an annular inner shell disposed inside of said annular outer shell, and an annular cavity between said annular inner shell and an annular outer shell. The annular cavity can be configured to receive the hollow base portion (61).

The implant is shown in Figure 2.20.

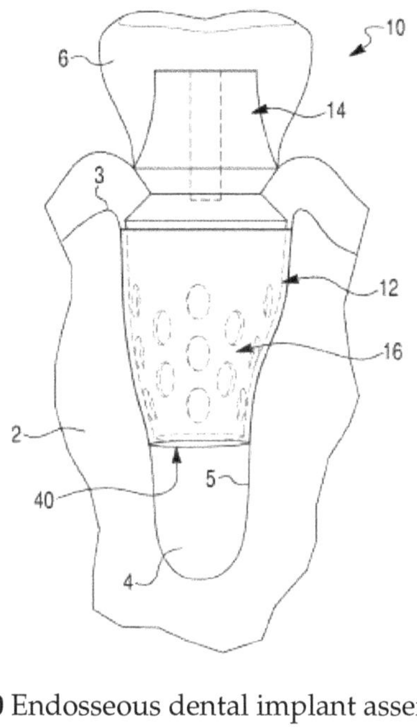

Figure 2.20 Endosseous dental implant assembly (61).

The dental implant assembly 10 in Figure 2.20 consists of an endosseous dental implant 12, and a bio-supportive or biodegradable scaffold 40 mounted to the dental implant 12. The dental implant 12 consists of an abutment portion 14 and a hollow base (or basket) portion 16 formed integrally with the abutment portion 14. The endosseous dental implant 12 is in the form of a one-piece body consisting of the abutment portion 14 and the hollow base portion 16. Alternatively, the abutment portion 14 and the hollow base portion 16 may be formed separately, then non-moveably secured to each other. Conventionally, the abutment portion 14 of the dental implant 12 is provided for engaging a complementary transfer component, comfort cap, or for eventually supporting a tooth crown 6 (61).

The base portion 16 resembles an inverted hollow basket or socket frame. The outer peripheral surface of the base portion 16 is a rough surface so as to have a predetermined external roughness to enhance bone contact in the osseointegration approach or to enhance stem cell growth in the biomimetic approach. In other words, the outer peripheral surface 22 of the base portion 16 is treated to enhance roughness of the base portion 16 (61).

2.14.4 *Titanium-Polymer Composites*

Titanium alloys are appropriate metals for biomedical applications such as implants (62). However, the great difference between the stiffness of the Ti alloys and compact bone results in stress shielding of the bone and stress concentration at the implant. Both these properties are undesirable and could result in a failure of the implant.

A method to reduce the stiffness of dense implants and to avoid the stress shielding is to add porosity to the structure. However, this results in a considerable reduction in the toughness of the structure, which is undesirable for the long-term success of implants.

A method for independently tuning the stiffness and toughness of the material has been developed by the addition of various polymers to the additively manufactured porous Ti structures (62).

Porous Ti samples with two levels of porosity were fabricated through additive manufacturing. Three different types of thermoplastic polymers were used to fill the pores and to make the titanium-polymer composite parts.

Compression simulations and tests were performed with both porous and composite specimens to compare the mechanical behavior of these structures. The results of simulation and the experimental findings indicated that filling porous Ti with thermoplastic polymers leads to an increase in the toughness of the structure. The percentage increase of the toughness was assessed as a function of the polymer toughness and the level of porosity in various samples. The results indicated a way for designing polymer-composite structures with independently tunable stiffness and toughness (62).

2.14.5 *Titanium Implant Functionalization with Phosphate-Containing Polymers*

In a clinical study, titanium implants were functionalized with poly-(phosphoric acid) and phosphorylated pullulan (63).

Pullulan, cf., Figure 2.21, is a polysaccharide polymer consisting of maltotriose units. Pullulan is also known as α-1,4-α-1,6-glucan. Pullulan is a natural polymer produced commercially by a yeast-like fungus *Aureobasidium pullulans*. It is nontoxic, non-immunogenic, non-carcinogenic and non-mutagenic in nature (64).

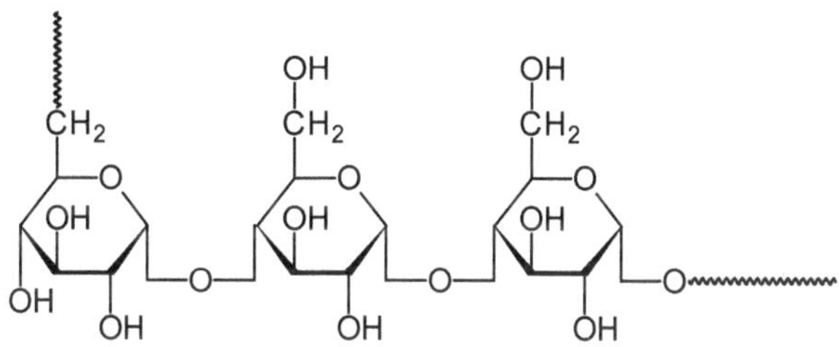

Figure 2.21 Pullulan.

The implant osseointegration was evaluated by quantitative histology, i.e., peri-implant bone formation, and the bone-to-implant contact.

It was found that the functionalized titanium implants with inorganic or organic phosphate-containing polymers at a concentration of 10% may stimulate the peri-implant bone formation and implant osseointegration so as to result in early healing times (63).

2.14.6 *Zirconia Dental Implants*

Zirconium dioxide, i.e., zirconia, has earned a place as a potential substitute for titanium in implantology (65). Despite the excellent biocompatibility and tissue integration of zirconia, low affinity to plaque and favorable biomechanical properties, certain early failures were significantly higher for zirconia implants than for titanium implants.

Technical failure as a result of fracture of the material has also been a major concern. So far, zirconia implants have mainly been manufactured as one-piece implant systems because of the material's limitations. Nevertheless, various two-piece systems have been progressively emerging with promising results. Also, screw-retained abutments are desirable but they have major technical challenges. Certain innovations and technical advances would lead to a further improvement in the reliability and strength of zirconia implants, thus allowing novel designs, connections and reconstructions (65).

The recent issues of the application of zirconia for dental implants have been reviewed (65).

Zirconium dioxide (zirconia) ceramics with improved properties have been introduced as an alternative material to aluminium oxide. They were first used for the fabrication of crowns and implant abutment (66,67),

Currently, tetragonal zirconia polycrystal, in particular 3 mol% yttrium oxide-stabilized zirconia, is the ceramic of choice for dental implants (68). The white, opaque color of zirconia, and the good biocompatibility and low affinity to bacterial plaque, make this material of interest in biomedical sciences.

Also, zirconia exhibits some promising physical and mechanical properties, including a low thermal conductivity, a high flexural strength of 900–1,200 *MPa*, favorable fracture resistance, and good wear and corrosion resistance (65).

Phase transformation toughening gives zirconia its excellent properties (69). This phenomenon stops the crack propagation resulting from the transformation of zirconia from the tetragonal phase into the monoclinic phase and the consequent 4% volume expansion and induction of compressive stresses.

The low-temperature degradation and aging behavior are negative properties of zirconia. In the presence of water or water vapor, a slow transformation from the tetragonal phase into the monoclinic phase takes place, which leads to a slow development of roughness, thus producing progressive deterioration of the material (70).

Aging occurs as a result of compressive stresses and microcracking. The degree of aging is dependent on the balance between these two factors. Aging may be influenced by various aspects of the production process, such as the macroscopic shape and the surface characteristics of an implant (65).

A significantly higher content of collagen and a shorter length of the sulcular epithelium were observed around zirconia implants of 0.76 *mm*, compared with 1.4 *mm* around titanium implants. The biological width was 2.3 *mm* for titanium implants and 2.85 *mm* for zirconia implants. It was hypothesized that the longer junctional epithelium and the higher density of collagen fibers could improve the soft-tissue seal and reduce the inflammatory infiltration around zirconia implants. Therefore, zirconia implants could result in a somewhat more mature soft-tissue integration. Figure 2.22 shows the soft-tissue healing 3 months after placement of a two-piece zirconia implant (65,71).

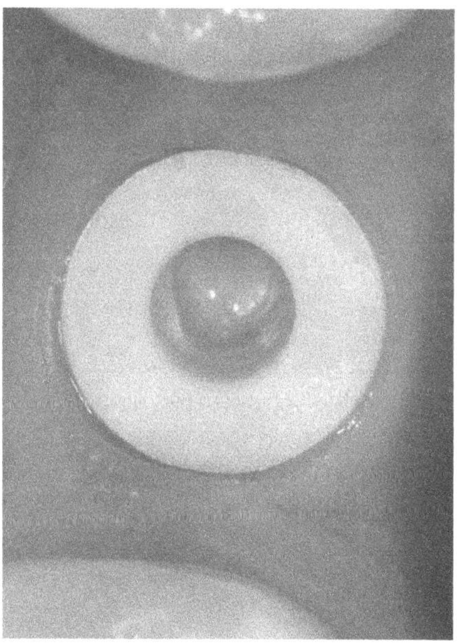

Figure 2.22 Soft-tissue healing 3 months after implant placement of a two-piece zirconia implant. Reprinted from (65,71) with permission from Wiley.

Clinical pictures and radiographs of a premolar replaced with a two-piece zirconia system, at 1 and 4 years after loading, are shown in Figure 2.23.

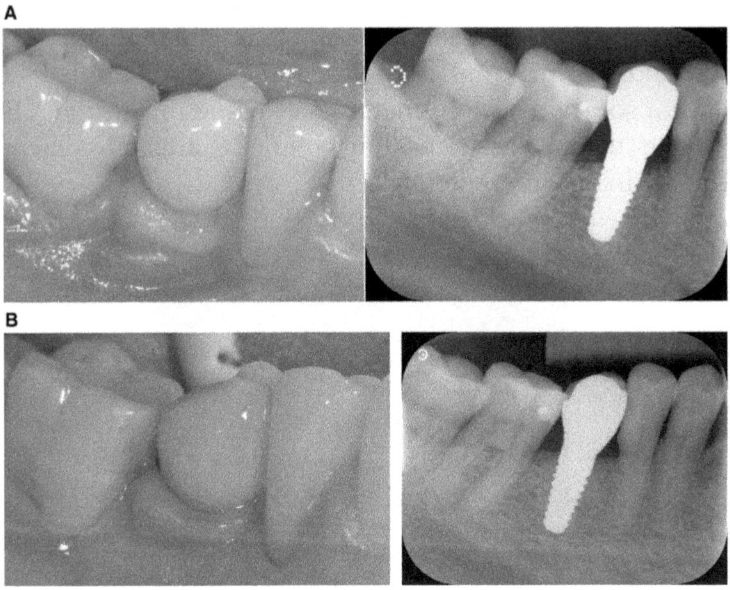

Figure 2.23 (A) Clinical photograph (left image) and radiograph (right image), 1 year after loading (two-piece zirconia implant). (B) Clinical photograph (left image) and radiograph (right image), 4 years after loading (two-piece zirconia implant) (65,71). Reprinted from (65,71) with permission from Wiley.

2.14.7 *Shape-Memory Polymers*

In restorative dentistry, implantable dental devices are used to achieve reinforcement of damaged or diseased teeth and permanent replacement of missing teeth that may have been lost or need to be replaced due to trauma, decay or disease. Dental implants can be implanted into the root canal space of a tooth to reinforce existing tooth structure and can be implanted into the root canal space of a tooth or a cavity within alveolar bone to provide means for fixation of a dental prosthesis such as a crown or bridge.

An implantable dental device has been described with a polymeric shape-memory material for implantation into a cavity within alveolar bone of the jaw or within the root canal space of a tooth (72).

Shape-memory alloys, such as Ni-Ti alloys, have been suggested as enhanced fixation materials for forming dental implants. These materials, upon induction with heat, are able to expand to a preconfigured shape, enabling enhanced implant fixation.

However, shape-memory alloys are expensive. Furthermore, the use of shape-memory alloys in dental implants can lead to problems associated with non-degradability and a lack of biocompatibility. For example, alloys containing nickel can initiate an allergic response in some cases.

As set out above, the use of shape-memory polymers in the field of dental implants has been extremely limited and has not been used in an implant that addresses these needs. It has now been determined that the provision of an implantable dental device which comprises shape-memory polymer arranged to directly contact the walls of an implant cavity, allows instant fixation with a simple implantation procedure and enhanced long-term fixation to be achieved.

The use of a polymeric shape-memory material provides a degree of flexibility with regard to the size of cavity an implant can be fixated within (72).

Shape-memory polymers are active polymers that have a dual-shape capability (73,74). They can change their shape in a predefined way from their original permanent shape to a deformed state (temporary shape) when exposed to an appropriate stimulus. The fundamental aspects of the molecular design of shape-memory polymers have been reviewed (73). Besides their dual-shape capability, such active materials can be biofunctional or biodegradable (73).

If a resorbable polymeric shape-memory material is used, over time as osseointegration occurs the material will be resorbed and replaced by newly formed bone (72). The use of resorbable shape-memory material imparts the device with osteoconductive properties, allowing, over time, the implantation cavity to be populated with regenerated bone cells. This is desirable for long-term integration of the device.

Shape-memory polymers, which can be resorbable or non-resorbable, are summarized in Table 2.10.

Table 2.10 Shape-memory polymers (72).

Polymer
Poly(ether ether ketone) (PEEK)
Poly(methyl methacrylate) (PMMA)
Poly(ethyl methacrylate) (PEMA)
Polyacrylate
Poly-α-hydroxy acids
Polycapropactones
Polydioxanones
Polyesters
Polyglycolic acid
Polyglycols
Polylactides
Polyorthoesters
Polyphosphates
Polyoxaesters
Polyphosphoesters
Polyphosphonates
Polysaccharides
Polytyrosine carbonates
Poly(urethane)s

Preferably, the polymeric shape-memory material resides in a deformed state below the glass transition temperature (T_g) and is activatable from the deformed state to the relaxed state above this temperature. Generally, polymeric materials that display shape-memory properties show a large change in modulus of elasticity at the glass transition temperature (T_g). Shape-memory properties are utilized by taking advantage of this characteristic. Namely, a macroscopic body of polymeric shape-memory material to which a definite shape (the original shape) has been imparted by molding, can be softened by providing the article with energy and heating to a temperature (T_f) higher than the T_g of the polymeric material, but lower than the melting temperature (T_m). At this temperature (T_f), the material can be deformed into a different macroscopic shape (the deformed state). In the deformed state an oriented polymer network is formed. The polymeric material is then cooled to a temperature lower than the T_g, whilst maintaining its deformed state. When the polymeric material is heated again to a temperature higher than the secondary molding temperature T_f, but lower than the T_m, the deformed state disappears and the polymeric material relaxes to recover its original shape. The input of energy necessary to cause the polymeric material to relax from its deformation state to its relaxed state is known as activation.

A method of implantation of an implantable dental device into a cavity within alveolar bone of the jaw has been described, wherein the method consists of the following steps (72):

1. Drilling a cavity in alveolar bone of the jaw,
2. Inserting the second end of the root member into the cavity, and
3. Activating the polymeric shape-memory portions of the root member to cause relaxation thereof, thereby fixating the root member within the cavity.

2.14.8 *Tetracycline Polymer Nanofiber Modified Titanium Disks*

The antimicrobial and osteogenic properties of titanium disks superficially modified with tetracycline-incorporated polymer nanofibers were investigated (75,76). Tetracycline is shown in Figure 2.24.

First, the synthesis and characterization, i.e., morphology, mechanical strength, drug release, antimicrobial activity, and cytocompatibility of tetracycline-incorporated fibers were done.

Then, both the antimicrobial and a murine-derived osteoprecursor cell response of titanium modified with tetracycline-incorporated fibers were evaluated. Tetracycline could be successfully incorporated into the submicron-sized and cytocompatible fibers (75).

All tetracycline-incorporated mats showed a significant antimicrobial activity against periodontal pathogens. The antimicrobial

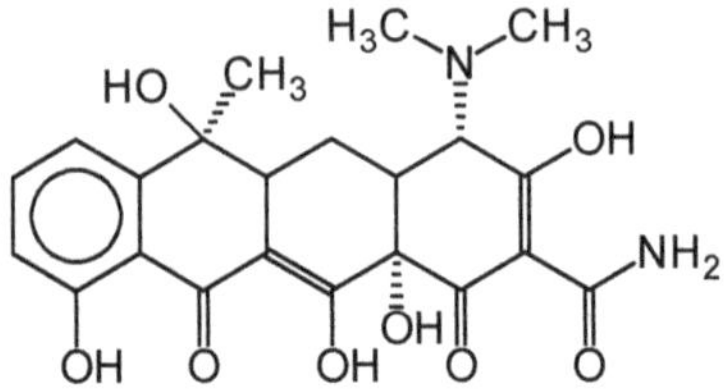

Figure 2.24 Tetracycline.

potential of the tetracycline-incorporated fiber modified titanium was influenced by both the tetracycline concentration and the tested bacteria. After 5 *d* and 7 *d*, a significant increase of the murine-derived osteoprecursor cell numbers was observed for tetracycline-incorporated nanofiber modified titanium disks in comparison to that of tetracycline free nanofiber modified titanium disks and bare titanium.

Also, a significant increase in the alkaline phosphatase levels on the titanium disks modified with tetracycline-incorporated nanofiber on days 7 and 14 was observed. This suggests that the proposed surface promotes early osteogenic differentiation. In summary, the results suggest that tetracycline-incorporated nanofibers could function as an antimicrobial surface modifier and osteogenic inducer for titanium dental implants (75).

2.14.9 *Biopolymers*

There are monographs concerning the issues of oral biomaterials (77). The recent developments in tissue engineering have been reviewed (78). In particular, repair mechanisms, the advantages of protein and polysaccharide derivatives, and the potential of stem cell therapy have been discussed. Also, the regenerative capacity of natural polymers used for the treatment of dentoalveolar defects have been described.

Tissue reconstruction using bioengineered scaffolds is more advantageous than the traditional autografting method, since it prevents the instigation of pain and donor site morbidity, while it ultimately can create both the environment and machinery needed to induce cell proliferation, migration, and reattachment within the affected area (78).

To overcome the drawbacks of commonly employed grafting materials, bone graft substitutes, such as 3D scaffolds, have been investigated in the dental field (79). Different biomaterials suitable for 3D scaffold fabrication have been reviewed (79). The review focuses on 3D-printed biomaterials as bone graft substitutes that might be convenient for various applications related to implant therapy. Also, their possible adoption for periodontal regeneration has been discussed.

2.14.9.1 Proteins

Natural polymers can be used, which include proteins and polysaccharides. These materials have high biocompatibility, good cell recognition, enhanced cellular interactions in the surrounding environment, and hydrophilicity.

Collagen is a protein in the human body which provides strength and structural stability to many tissues from skin to bone (79). Collagen matrices promote cell adhesion, proliferation, and osteogenic differentiation of bone marrow stromal cells.

Chitosan is a popular biomaterial in bone tissue engineering. It shows antibacterial and antifungal activity, rapid blood clot formation, and analgesic properties (79). These properties make chitosan useful in the acceleration of wound healing, which would minimize the risk of scaffold contamination and postoperative infections. Thus, eventual exposure and failure of the scaffold is prevented. For the same applications, alginate can be used. Alginate is highly processable for different scaffold types, which encourages its employment in regenerative medicine. Actually, alginate and chitosan do not exist within the human body, but they display structural similarities to glycosaminoglycans that are found in the extracellular matrix of human tissues such as bone.

Despite their good biological properties, the previously mentioned natural polymers lack bioactivity, which is the key factor in promoting hard tissue formation. They also have weak mechanical characteristics and a somewhat rapid degradation rate through enzymatic reactions. To overcome these undesired properties, scaffolds based on natural polymers are usually combined with bioactive materials, e.g., bioceramics, or mechanically strong materials, e.g., synthetic polymers or metals.

2.14.9.2 *Agarose*

Agarose is a biopolymer of polysaccharide chains that can be utilized as a scaffold in tissue engineering (80).

Agarose is a polysaccharide that is commonly extracted from seaweed (81). It is a linear polymer made up of the repeating unit of agarobiose, which is a disaccharide made up of *D*-galactose and 3,6-anhydro-*L*-galactopyranose.

2.14.9.3 *Gelatin*

Gelatin is a mixture of peptides and proteins obtained from the partial hydrolysis of natural animal collagen. This material is commercially utilized as a biodegradable polymer (82). Carrageenan is a polysaccharide polymer extracted from red seaweed and is utilized as a gelation agent or drug stabilizer (83). Xanthan gum is obtained via the fermentation of carbohydrates with *Xanthomonas campestris* and is often utilized to control the viscosity of food (84). Alginate is a characteristic polysaccharide that forms a hydrogel in the presence of divalent metal ions and can be utilized as both wound dressing and scaffolds (85). Pectins are extracted from the peels of citrus fruits and are widely used as food stabilizers (86). Hyaluronic acid is used in implant coatings or during ocular surgery due to its high water content in its hydrogel form (87).

2.14.9.4 *Biodegradable Synthetic Polymers*

Biodegradable synthetic polymers are of interest because of their relatively low cost and ability to be produced in large quantities with long shelf life in comparison to their natural counterparts (79). Biodegradable synthetic polymers are poly(caprolactone) (PCL), poly(lactic acid) (PLA), poly(glycolic acid) (PGA), and their copolymer poly(lactic-*co*-glycolic acid) (PLGA).

Due to its biocompatibility, PCL is an excellent candidate for bone tissue engineering applications, and it is suitable for various scaffold fabrication techniques. It has a remarkably slow degradation rate, and good mechanical stability. However, PCL is hydrophobic in nature, which is responsible for its inferior cell affinity and poor cellular responses and interactions to the surface.

Similar to PCL, PLA and PLGA are hydrophobic. In contrast, PGA is hydrophilic. But these polymers have higher rates of degradation in comparison to PCL. However, in general, aliphatic polyesters show a slow degradation rate in comparison to natural polymers and bioceramics. In general, synthetic polymers degrade by hydrolysis, which can occur in the form of bulk degradation or surface erosion (79).

2.14.9.5 Bioceramics

Bioceramics are inorganic biomaterials of different categories. Examples are calcium phosphate bioceramics and bioactive glass with very well-documented applications as bone fillers in the dental field. Calcium phosphate bioceramics may contain hydroxyapatite, tricalcium phosphate, and biphasic calcium phosphate.

These materials may be in the form of injectable cement materials, i.e., pastes that are moldable and easy to handle and harden when left *in-situ*. Moldable calcium phosphate materials allow for an intimate adaptation to complex defects, which is difficult to accomplish with conventional bone grafting materials (79).

Another biomaterial that belongs to bioceramics is bioactive glass, which is a silicon oxide with substituted calcium. When exposed to body fluids, a layer of calcium phosphate is formed on the surface of the bioactive glass, which chemically binds to the bone.

A disadvantage of bioceramics is that these materials are extremely brittle and difficult to shape into the desired structures because of their stiffness and low flexibility and moldability. Also, they exhibit a weak mechanical strength and fracture toughness. These properties limit their applications to non-load-bearing areas. However, their combination with mechanically strong biomaterials, such as synthetic polyesters or metals, may eliminate the brittleness, the difficulty in shaping, and the weakness in mechanical strength (79).

2.14.9.6 Metallic Biomaterials

Metallic biomaterials have been extensively used in dental and orthopedic fields to support the replacement of lost bone structures (79).

They have a high strength, toughness, and hardness in comparison to polymers and ceramics. These properties make them suitable for applications in load-bearing areas. Metals enhance the mechanical properties of a scaffold by decreasing the pore size.

Within this group of biomaterials, titanium and titanium alloys are encouraged in bone regeneration due to their high biocompatibility, appropriate mechanical properties, and elasticity. Titanium-based 3D scaffolds display a good hydrophilicity, which enhances mineral deposition and encourages cell attachment and proliferation. Also, new bone formation without any signs of inflammation or necrosis have been found *in-vivo* (79).

However, the lack of biodegradability of titanium and titanium alloys is a major disadvantage and might discourage their applications in bone regeneration due to the need for a second surgery for removal, which can compromise patient satisfaction and increase health care costs.

References

1. D.-G. Kim, Hybrid dental implant, US Patent 9 757 213, assigned to Ohio State Innovation Foundation (Columbus, OH), September 12, 2017.
2. C.Q. Wang, Method of dental implant restoration, US Patent 9 763 753, assigned to Zuga Medical, Inc. (Beachwood, OH), September 19, 2017.
3. A.P. Saadoun, *Esthetic implant treatment* in *Esthetic Soft Tissue Management of Teeth and Implants*, chapter 6. John Wiley & Sons, Ltd., West Sussex, UK, 2012.
4. R.W. Towse, M.D. Scalise, and Z.B. Suttin, Dental implant system having enhanced soft-tissue growth features, US Patent 9 168 110, assigned to Biomet 3I, LLC (Palm Beach Gardens, FL), October 27, 2015.
5. D.P. Rogers, G.R. Brown, R.E. Goodman, and R.J. Lazzara, Single-stage implant system, US Patent 6 394 809, assigned to Implant Innovations, Inc. (Palm Beach Gardens, FL), May 28, 2002.
6. W. Groll, D. Wolf, A. Spanel, T. Lange, and N. Bergner, Dental implant component, US Patent Application 20 060 246 398, assigned to Friadent Gmbh (Mannheim DE), November 2, 2006.
7. S.J. Lee, M.S. Bae, D.W. Lee, D.N. Heo, D. Lee, M. Heo, S.-J. Hong, J. Kim, W.D. Kim, S.A. Park, and I.K. Kwon, *Carbohydrate Polymers*, Vol. 157, p. 1750, 2017.

8. G.-H. Gwak, A.-J. Choi, Y.-S. Bae, H.-J. Choi, and J.-M. Oh, *Biomaterials Research*, Vol. 20, p. 1, Feb 2016.
9. Y.S. Choi, S. Lee, S.R. Hong, Y. Lee, K. Song, and M. Park, *Journal of Materials Science: Materials in Medicine*, Vol. 12, p. 67, 2001.
10. G.-H. Gwak, S.-M. Paek, and J.-M. Oh, *European Journal of Inorganic Chemistry*, Vol. 2012, p. 5269, 2012.
11. P. Proussaefs and A. AlHelal, *The Journal of Prosthetic Dentistry*, Vol. in press, p. in press, 2017.
12. C.F. Andreescu, D.L. Ghergic, O. Botoaca, H.M. Barbu, I.S.C. Mitariu, and D.N. Patroi, *Materiale Plastice*, Vol. 54, p. 32, 2017.
13. B. Stawarczyk, I. Sailer, B. Sapina, A. Ender, A. Trottmann, and C. Hämmerle, *Dental Dialogue*, Vol. 10, p. 30, 2009.
14. J.-F. Güth, J.S. Almeida E. Silva, M. Ramberger, F. Beuer, and D. Edelhoff, *Journal of Esthetic and Restorative Dentistry*, Vol. 24, p. 310, 2012.
15. C. Grenade, M.-C. De Pauw-Gillet, P. Gailly, A. Vanheusden, and A. Mainjot, *Dental Materials*, Vol. 32, p. 1152, 2016.
16. C. Grenade, M.-C. De Pauw-Gillet, C. Pirard, V. Bertrand, C. Charlier, A. Vanheusden, and A. Mainjot, *Dental Materials*, Vol. 33, p. 333, 2017.
17. T.J. Ferreira, M.T. Vieira, J. Costa, M. Silva, and P.T. Gago, *Journal of Orthodontics & Endodontics*, Vol. 2, 2016.
18. F.M. Barreiros, M.-T. Vieira, and J. Castanho, *Metal Powder Report*, Vol. 64, p. 18, 2009.
19. K.S. Hassan, *Oral Surgery, Oral Medicine, Oral Pathology, Oral Radiology, and Endodontology*, Vol. 108, p. e19, 2009.
20. L. Biasetto and H. Elsayed, *Surface and Coatings Technology*, Vol. 321, p. 416, 2017.
21. L. Biasetto, H. Elsayed, F. Bonollo, and P. Colombo, *Surface and Coatings Technology*, Vol. 301, p. 140, 2016.
22. J. Howard, S. Patel, D.K. Mills, and S. Gold, Assessment of cell-material interactions on 3D nanostructured titania-polymer surfaces towards the improvement of osseointegration of orthopedic and dental implants, in *The 2008 Annual Meeting*, Philadelphia, PA, 2008.
23. K. Das, S. Bose, and A. Bandyopadhyay, *Acta Biomaterialia*, Vol. 3, p. 573, 2007.
24. T.K. Monsees, K. Barth, S. Tippelt, K. Heidel, A. Gorbunov, W. Pompe, and R.H.W. Funk, *Cells Tissues Organs*, Vol. 180, p. 81, 2005.
25. O. Moses, O. Zigman, S. Slutzki, O. Cohen, H. Tal1, and S. Matalon, Temperature changes in one-piece implants due to provisional restoration, in *Abstracts of the EAO Congress, Paris, France, 29 September – 1 October 2016*, Vol. 27 of *Clinical Oral Implants Research*, p. 86, 2016.
26. P.S. Lyren, Method for forming a dental implant with porous body, US Patent 9 271 811, March 1, 2016.

27. W.R. Wagner and B. Story, Dental implant having multiple textured surfaces, US Patent 5 989 027, assigned to Sulzer Calcitek Inc. (Carlsbad, CA), November 23, 1999.
28. Y. Daigo, S. Ohmori, and K. Kato, Porous implant material, US Patent 9 707 321, assigned to Mitsubishi Materials Corporation (Tokyo, JP), July 18, 2017.
29. S.D. Jensen, Dental implant utilizing a polymeric post, US Patent 9 211 167, assigned to CAO Group, Inc. (West Jordan, UT), December 15, 2015.
30. C.A.A. Lemos, M.L. Ferro-Alves, R. Okamoto, M.R. Mendonça, and E.P. Pellizzer, *Journal of Dentistry*, Vol. 47, p. 8, 2016.
31. P. Yun, T. Tsai, B. Yun, and T. Lee, Adjustable dental implant, US Patent 9 775 689, assigned to Geo-Protector Tech. Co., Ltd. (Kaohsiung, TW), October 3, 2017.
32. S. Najeeb, M.S. Zafar, Z. Khurshid, and F. Siddiqui, *Journal of Prosthodontic Research*, Vol. 60, p. 12, 2016.
33. M. Al-Rabab'ah, W. Hamadneh, I. Alsalem, A. Khraisat, and A. Abu Karaky, *Journal of Prosthodontics*, Vol. in press, p. n/a, 2017.
34. A.D. Schwitalla, T. Spintig, I. Kallage, and W.-D. Müller, *Dental Materials*, Vol. 31, p. 1377, 2015.
35. Technical Committee: ISO/TC 106/SC 2, Dentistry — polymer-based crown and bridge materials, ISO Standard ISO 10477, International Organization for Standardization, Geneva, Switzerland, 2004.
36. M. Zhao, M. An, Q. Wang, X. Liu, W. Lai, X. Zhao, S. Wei, and J. Ji, *Journal of Proteomics*, Vol. 75, p. 3560, 2012.
37. F.P. Koch, D. Weng, S. Krämer, S. Biesterfeld, A. Jahn-Eimermacher, and W. Wagner, *Clinical Oral Implants Research*, Vol. 21, p. 350, 2010.
38. A. Schwitalla and W.-D. Müller, *Journal of Oral Implantology*, Vol. 39, p. 743, 2013.
39. R.F.V. Rocha, L.C. Anami, T.M.B. Campos, R.M. de Melo, R.O. de Assunção e Souza, and M.A. Bottino, *Brazilian Dental Journal*, Vol. 27, p. 693, November 2016.
40. A.D. Schwitalla, T. Spintig, I. Kallage, and W.-D. Müller, *Journal of the Mechanical Behavior of Biomedical Materials*, Vol. 54, p. 295, 2016.
41. A.D. Schwitalla, T. Zimmermann, T. Spintig, I. Kallage, and W.-D. Müller, *Journal of the Mechanical Behavior of Biomedical Materials*, Vol. 69, p. 163, 2017.
42. M. Sampaio, M. Buciumeanu, B. Henriques, F. Silva, J. Souza, and J. Gomes, *Journal of the Mechanical Behavior of Biomedical Materials*, Vol. 60, p. 212, 2016.
43. M. Sampaio, M. Buciumeanu, B. Henriques, F.S. Silva, J.C. Souza, and J.R. Gomes, *Journal of the Mechanical Behavior of Biomedical Materials*, Vol. 54, p. 123, 2016.

44. P. Zoidis and I. Papathanasiou, *The Journal of Prosthetic Dentistry*, Vol. 116, p. 637, 2016.
45. A.D. Schwitalla, M. Abou-Emara, T. Zimmermann, T. Spintig, F. Beuer, J. Lackmann, and W.-D. Müller, *Journal of the Mechanical Behavior of Biomedical Materials*, Vol. 63, p. 244, 2016.
46. Camlog, CAMLOG implant systems, electronic: http://www.camlog.com/en/, 2017.
47. F. Suska, O. Omar, L. Emanuelsson, M. Taylor, P. Gruner, A. Kinbrum, D. Hunt, T. Hunt, A. Taylor, and A. Palmquist, *Journal of Biomaterials Applications*, Vol. 29, p. 234, 2014.
48. S. Barkarmo, A. Wennerberg, M. Hoffman, P. Kjellin, K. Breding, P. Handa, and V. Stenport, *Journal of Biomedical Materials Research Part A*, Vol. 101A, p. 465, 2013.
49. S. Barkarmo, M. Andersson, F. Currie, P. Kjellin, R. Jimbo, C. Johansson, and V. Stenport, *Clinical Oral Implants Research*, Vol. 25, p. 178, 2014.
50. P. Johansson, R. Jimbo, P. Kjellin, F. Currie, B.R. Chrcanovic, and A. Wennerberg, *International Journal of Nanomedicine*, Vol. 9, p. 3903, August 2014.
51. D. Vogel, H. Dempwolf, A. Baumann, and R. Bader, *Journal of the Mechanical Behavior of Biomedical Materials*, Vol. in press, 2017.
52. J. Waser-Althaus, A. Salamon, M. Waser, C. Padeste, M. Kreutzer, U. Pieles, B. Müller, and K. Peters, *Journal of Materials Science: Materials in Medicine*, Vol. 25, p. 515, Feb 2014.
53. C.-M. Han, E.-J. Lee, H.-E. Kim, Y.-H. Koh, K.N. Kim, Y. Ha, and S.-U. Kuh, *Biomaterials*, Vol. 31, p. 3465 , 2010.
54. Y. Zhao, H.M. Wong, W. Wang, P. Li, Z. Xu, E.Y. Chong, C.H. Yan, K.W. Yeung, and P.K. Chu, *Biomaterials*, Vol. 34, p. 9264, 2013.
55. M. Pino, N. Stingelin, and K.E. Tanner, *Acta Biomaterialia*, Vol. 4, p. 1827, 2008.
56. H. Zhou, V.K. Goel, and S.B. Bhaduri, *Materials Letters*, Vol. 125, p. 96 , 2014.
57. M.S.A. Bakar, M.H.W. Cheng, S.M. Tang, S.C. Yu, K. Liao, C.T. Tan, K.A. Khor, and P. Cheang, *Biomaterials*, Vol. 24, p. 2245, 2003. Focus on Biomaterials Science in Asia.
58. L. Wang, S. He, X. Wu, S. Liang, Z. Mu, J. Wei, F. Deng, Y. Deng, and S. Wei, *Biomaterials*, Vol. 35, p. 6758, 2014.
59. A. Xu, L. Zhou, Y. Deng, X. Chen, X. Xiong, F. Deng, and S. Wei, *Journal of Materials Chemistry B*, Vol. 4, p. 1878, 2016.
60. D. McDevitt, Expandable polymer dental implant and method of use, US Patent 6 840 770, January 11, 2005.
61. B.B. Westover, Endosseous dental implant assembly, US Patent 9 730 771, assigned to Westover and Brock B. (Ridgeland, MS), August 15, 2017.

62. S.N. Esfahani, M.T. Andani, N.S. Moghaddam, R. Mirzaeifar, and M. Elahinia, *Journal of Materials Processing Technology*, Vol. 238, p. 22, 2016.
63. M.V. Cardoso, J. de Rycker, A. Chaudhari, E. Coutinho, Y. Yoshida, B. Van Meerbeek, M.F. Mesquita, W.J. da Silva, K. Yoshihara, K. Vandamme, and J. Duyck, *Journal of Clinical Periodontology*, Vol. 44, p. 950, 2017.
64. R.S. Singh, N. Kaur, V. Rana, and J.F. Kennedy, *Carbohydrate Polymers*, Vol. 171, p. 102, 2017.
65. N. Cionca, D. Hashim, and A. Mombelli, *Periodontology 2000*, Vol. 73, p. 241, 2017.
66. B. Al-Amleh, K. Lyons, and M. Swain, *Journal of Oral Rehabilitation*, Vol. 37, p. 641, 2010.
67. K. Nakamura, T. Kanno, P. Milleding, and U. Örtengren, *International Journal of Prosthodontics*, Vol. 23, 2010.
68. J.R. Kelly and I. Denry, *Dental Materials*, Vol. 24, p. 289, 2008.
69. C. Sanon, J. Chevalier, T. Douillard, M. Cattani-Lorente, S.S. Scherrer, and L. Gremillard, *Dental Materials*, Vol. 31, p. 15, 2015.
70. V. Lughi and V. Sergo, *Dental Materials*, Vol. 26, p. 807, 2010.
71. N. Cionca, N. Müller, and A. Mombelli, *Clinical Oral Implants Research*, Vol. 26, p. 413, 2015.
72. P. Arnold, D. Sarefjord, and A. Thompson, Dental implant, US Patent 9 545 291, assigned to Smith & Nephew PLC (London, GB), January 17, 2017.
73. M. Behl and A. Lendlein, *Materials Today*, Vol. 10, p. 20, 2007.
74. Wikipedia, Shape-memory polymer — wikipedia, the free encyclopedia, 2017. [Online; accessed 8-November-2017].
75. M.C. Bottino, E.A. Münchow, M.T.P. Albuquerque, K. Kamocki, R. Shahi, R.L. Gregory, T.-M.G. Chu, and D. Pankajakshan, *Journal of Biomedical Materials Research Part B: Applied Biomaterials*, Vol. 105, p. 2085, 2017.
76. R.G. Shahi, M.T.P. Albuquerque, E.A. Münchow, S.B. Blanchard, R.L. Gregory, and M.C. Bottino, *Odontology*, Vol. 105, p. 354, Jul 2017.
77. J. Matinlinna, *Handbook of Oral Biomaterials*, Pan Stanford Publishing, Singapore, 2014.
78. H.E. Jazayeri, M.D. Fahmy, M. Razavi, B.E. Stein, A. Nowman, R.M. Masri, and L. Tayebi, *Journal of Prosthodontics*, Vol. 25, p. 510, 2016.
79. F. Asa'ad, G. Pagni, S.P. Pilipchuk, A.B. Giannì, W.V. Giannobile, and G. Rasperini, *International Journal of Dentistry*, Vol. 2016, p. 1, 2016.
80. R.L. Mauck, M.A. Soltz, C.C.B. Wang, D.D. Wong, P.-H.G. Chao, W.B. Valhmu, C.T. Hung, and G.A. Ateshian, *Journal of Biomechanical Engineering*, Vol. 122, p. 252, 2000.

81. Wikipedia, Agarose — wikipedia, the free encyclopedia, 2017. [Online; accessed 9-December-2017].
82. T. Hayashi, *Progress in Polymer Science*, Vol. 19, p. 663, 1994.
83. V.L. Campo, D.F. Kawano, D.B. da Silva, and I. Carvalho, *Carbohydrate Polymers*, Vol. 77, p. 167, 2009.
84. C.A. Edwards, N.A. Blackburn, L. Craigen, P. Davison, J. Tomlin, K. Sugden, I.T. Johnson, and N.W. Read, *The American Journal of Clinical Nutrition*, Vol. 46, p. 72, 1987.
85. L. Wang, R. Shelton, P. Cooper, M. Lawson, J. Triffitt, and J. Barralet, *Biomaterials*, Vol. 24, p. 3475, 2003.
86. M. Tasneem, F. Siddique, A. Ahmad, and U. Farooq, *Critical Reviews in Food Science and Nutrition*, Vol. 54, p. 869, 2014.
87. E.J. Oh, K. Park, K.S. Kim, J. Kim, J.-A. Yang, J.-H. Kong, M.Y. Lee, A.S. Hoffman, and S.K. Hahn, *Journal of Controlled Release*, Vol. 141, p. 2, 2010.

3

Dentures

3.1 Properties

3.1.1 Plaque Index

The issues and variations of the plaque index have been given elsewhere (1). One of the earlier indices for determining the amount of plaque on the tooth is the Modified Navy Plaque Index (MNPI) (2).

An improvement known as the Rustogi modification of the Modified Navy Plaque Index (Rustogi MNPI) was developed nearly 20 years later (3). The Rustogi MNPI differs from MNPI in that the former increases the tooth scoring areas in both the interproximal (mesial and distal) and gumline regions.

A mean plaque index (MPI) is calculated by summing the total number of tooth areas classified as having plaque present divided by the total number of tooth areas scored, e.g., 16 teeth classified as having plaque out of 32 teeth scored would have an MPI of 0.5. The efficacy of plaque removal can then be determined by comparing the the MPI before and after the attempted plaque removal activity, e.g., brushing the teeth or rinsing the mouth.

Improvements in plaque removal can be achieved by modifications to the Rustogi modification of the Modified Navy Plaque Index (4).

The modification consists of dividing a tooth into nine separate regions (A to I) as depicted in Figure 3.1, visually examining each region for plaque and scoring each region to obtain a plaque index number.

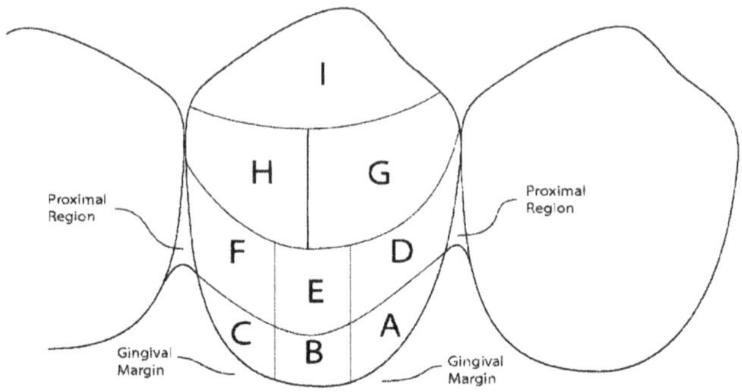

Figure 3.1 Separate regions of a tooth (4).

Each region is investigated for plaque by visual examination. A number is associated with every region: 0, if no plaque, 1, if the plaque covers less than one-third of the surface of the tooth, 2, if the plaque covers less than two-thirds of the surface of the tooth, 3, if the plaque cover two-thirds or more of the surface of the tooth. Then, the scores are totaled for each tooth examined to obtain a plaque index number (4).

3.1.2 *Inhibition of Denture Plaque Deposition*

The effects of treatment with a 2-methacryloyloxyethyl phosphorylcholine polymer, on plaque deposition in complete dentures were evaluated in a clinical study (5).

The mean percentage of plaque index of the polymer-treated dentures significantly decreased both on the mucosal surfaces and on the polished surfaces (5).

Removable dentures made of poly(methyl methacrylate) are prone to bacterial adherence and dental plaque formation. Denture plaque-associated infection is a source of serious dental and medical complications.

2-Methacryloyloxyethyl phosphorylcholine (MPC) is a well-known biomedical material that exhibits marked antithrombogenicity and tissue compatibility because of its high resistance to protein adsorption and cell adhesion. Therefore, MPC polymer coatings have been suggested to have the potential to inhibit plaque deposition on the surface of poly(methyl methacrylate) (PMMA) dentures.

However, coating MPC polymer on the surface of a PMMA denture is a complex procedure that requires specialized equipment, which is regarded as a major barrier to its clinical application.

A MPC polymer treatment procedure has been introduced that uses a copolymer that also contains *n*-butyl methacrylate (BMA) and 2-methacryloyloxyethyl 4-azidobenzoate (MBPAz), poly(MPC-*co*-BMA-*co*-MPAz) (PMBPAz) to prevent denture plaque deposition on removable dentures. The monomers are shown in Figure 3.2.

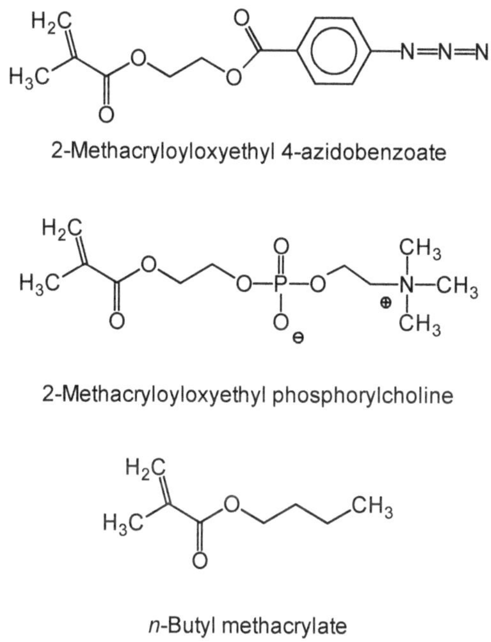

Figure 3.2 Comonomers to prevent denture plaque deposition (6).

This procedure enables the MPC coating of PMMA denture surfaces in a simple and stable manner that is resistant to various chemical and mechanical stresses due to the MPC layer of PMBPAz that is covalently bound to the PMMA surface by ultraviolet light irradiation. In addition, the procedure does not require any specialized equipment and can be completed by clinicians within 2 *min*. This procedure has been applied in a clinical setting and demonstrated its clinical utility and efficacy in inhibiting plaque deposition on removable dentures (6).

3.2 Materials

3.2.1 *Longevity of Fiber Reinforced Composite*

The clinical issues of longevity of fiber reinforced composite partial dentures have been reviewed (7).

3.2.2 *Denture Tooth and Material*

An improved dental tooth, and materials and methods for the preparation have been described (8).

A urethane monomer, an acidic monomer and one or more hydrophobic comonomers are used. The urethane group has the ability to hydrogen bond by acting as a hydrogen bond donor or hydrogen bond acceptor. The carboxylic acid group in methacrylic acid is capable of acting as a hydrogen bond acceptor or donor depending on the pH of the composition. Suitable monomers are collected in Table 3.1 and some are shown in Figure 3.3.

Also, a surfactant can be used in the composition. The surfactant can be selected from compounds shown in Table 3.2 and in Figure 3.4.

The resin composition can further contain a prepolymer that is formed from a resin composition comprising of methyl methacrylate or isobornyl methacrylate.

Barium glass, fumed silica, and ytterbium fluoride can be used as phase filler (8).

A shaped dental prosthetic device for use in a human mouth can be prepared by (8):

- Dispensing a urethane monomer-based resin composition and one or more hydrophobic monomers,
- Shaping the mixture into the form of the shaped dental prosthetic device, and
- Photopolymerizing the shaped mixture.

The photoinitiator can optionally further contain a coinitiator, for example, 2-ethyl hexylacrylate, or an amine coinitiator such as ethyl-4-(dimethylamino)benzoate, 2-ethylhexyl (dimethylamino)benzoate, or dimethylaminoethyl (meth)acrylate (8).

Table 3.1 Suitable monomers for an improved dental tooth (8).

Urethane monomers
1,6-Bis(Methacryloxy-2-ethoxycarbonylamino)-2,2,4-trimethylhexane
Bis(2-(methacryloyloxy)ethyl)-5,7,7,24,24,26-hexamethyl-10,21-dioxo-11,14,17,20-tetraoxa-2,9,22,29-tetraazatriacontanedioate
Acidic monomers
Methacrylic acid
Bis(2-methacryloxyethyl) phthalate
Pyromellitic glycerol dimethacrylate
Methacroyloxyethyl maleate
Hydroxyethyl methacrylate/succinate adduct
1,3-Glycerol dimethacrylate/maleate adduct
1,3-Glycerol dimethacrylate/succinate adduct
Hydrophobic monomers
Isostearyl methacrylate
Ethoxylated bisphenol A dimethacrylate
Stearyl methacrylate
Lauryl methacrylate
Isodecyl methacrylate
2-Ethylhexyl methacrylate
Cyclohexyl methacrylate

Table 3.2 Surfactants (8).

Compound
Sodium dodecyl sulfate
Ammonium lauryl sulfate
Cetyltrimethylammonium bromide
Cetylpyridinium chloride
Polyethoxylated tallow amine
Dodecyl betaine
Dodecyl dimethylamine oxide
Sodium lauryl sulfate
Polyether modified polydimethyl-siloxane (BYK®.-307)

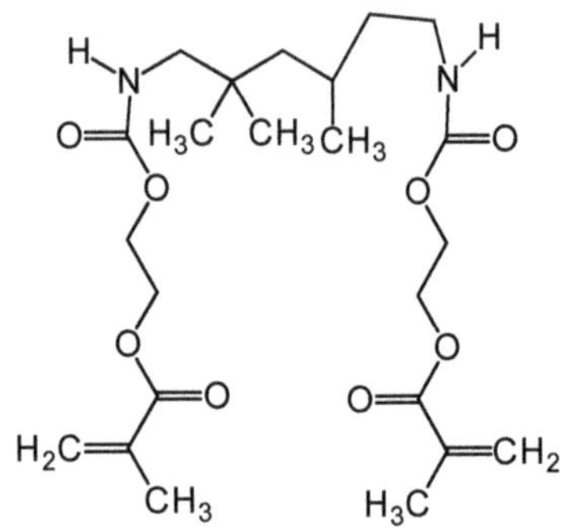

1,6-Bis(Methacryloxy-2-ethoxycarbonylamino)-2,2,4-trimethylhexane

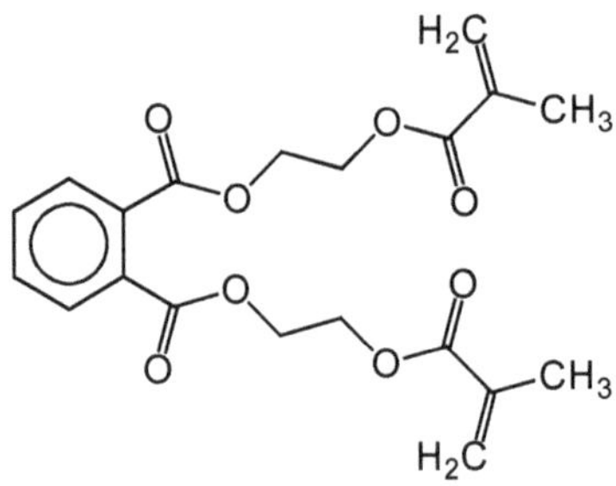

Bis (2-methacryloxyethyl) phthalate

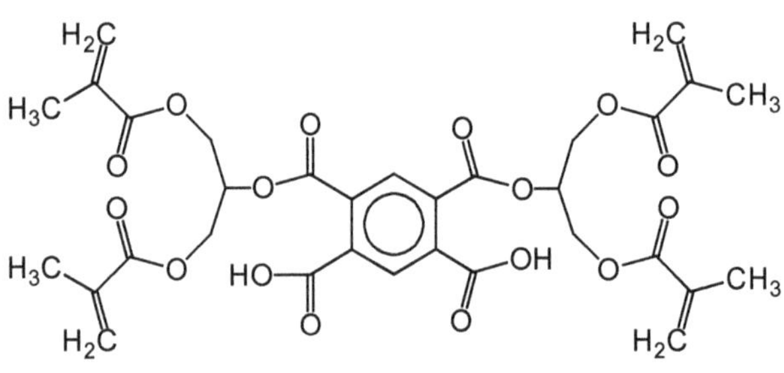

Pyromellitic glycerol dimethacrylate

Figure 3.3 Monomers.

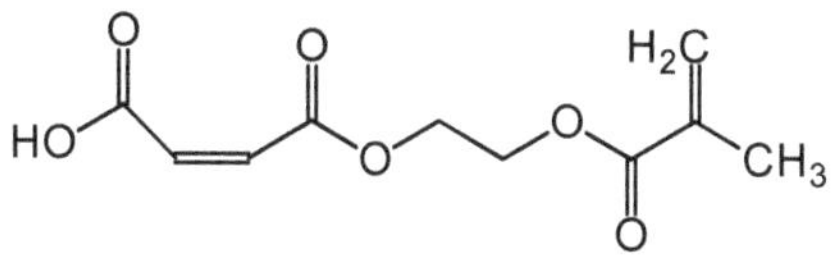

Methacroyloxyethyl maleate

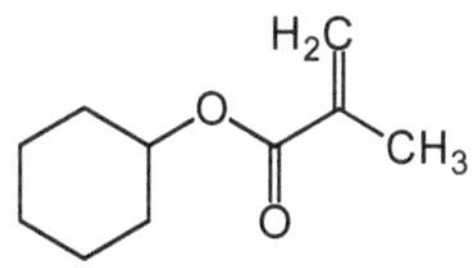

Cyclohexyl methacrylate

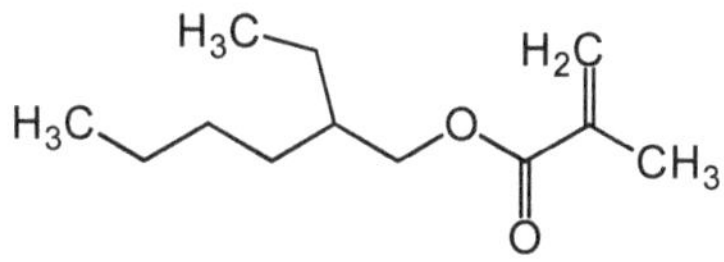

2-Ethylhexyl methacrylate

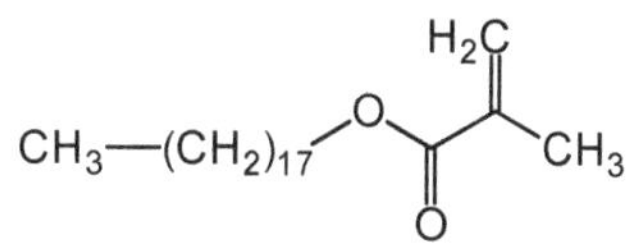

Stearyl methacrylate

Figure 3.3 (cont.) Monomers

3.2.3 Fixed Partial Dentures Made from Fiber Reinforced Polymer

Fiber reinforced polymers for the production of fixed partial dentures have been reviewed (9).

A fixed partial denture has been considered to be the best therapy for restoring missing teeth. However, a fixed partial denture is associated with potentially negative adverse effects, such as increased caries incidence and nonphysiologic loading of the retaining teeth due to a poor adaptation (9).

Various technological solutions for the fiber reinforcement of polymers have been used in commercial products. Such materials are shown in Table 3.3.

3.2.4 Denture Adhesives

Denture adhesives are used to assist a denture's ability to adhere to the gums (10). Denture adhesives are typically characterized as contributing an adhesive function once exposed to the moisture in a person's mouth and dissolve in the mouth over the course of a day.

A conventional approach is to use an adhesive material made of a resin in the form of a powder or paste, which can be either natural or synthetic. Each time a user needs to apply a denture to the gum, the user takes the required amount of a denture base stabilizing material from a tube in the case of a powder or paste. Conventional denture

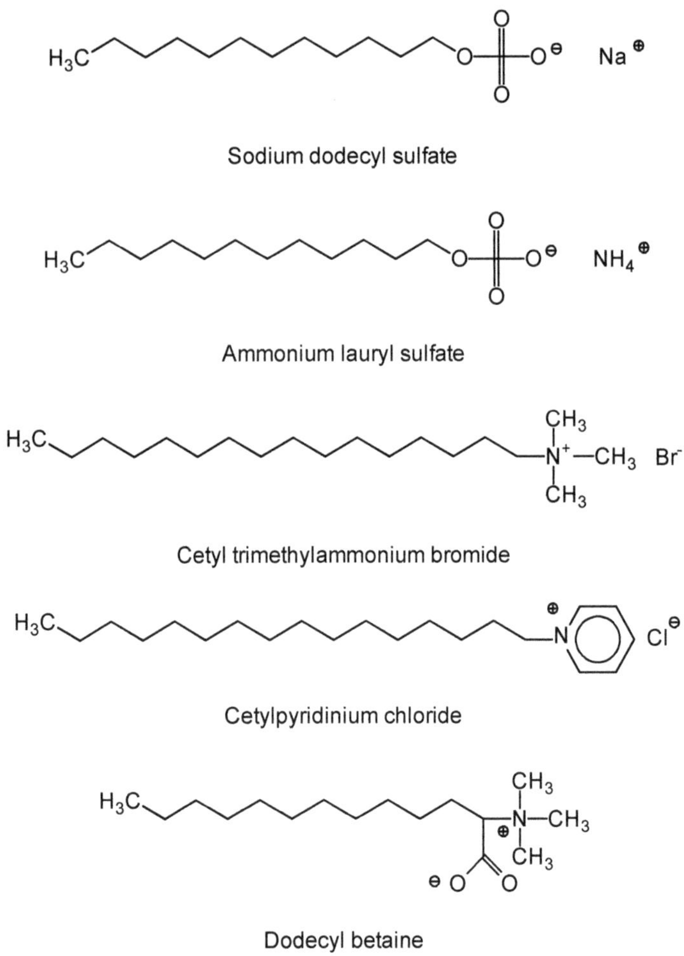

Figure 3.4 Surfactants (8).

Table 3.3 Alternatives to fixed partial dentures (9).

Material	Form
Poly(ethylene)	Woven fibers
Poly(ethylene)	Braided fibers
Poly(ethylene)	Unidirectional fibers
Poly(alkane)	Braided fibers
Kevlar	Unidirectional fibers
Glass	Woven fibers
Glass	Braided fibers
Glass	Unidirectional fibers
Glass	Unidirectional fibers in mesh

adhesives in the form of pastes and powders are highly hydrophilic and readily absorb moisture, forming an adhesive gel.

Such pastes and powders can be problematic for denture users. Once inserted, force is applied to the denture in order to spread the paste or power and ensure a snug fit of the denture to the gums. The force of application will cause the gums to displace the paste or gel that is formed and these adhesives will typically ooze or squirt out the side of the denture. This is known as primary ooze and is considered by many denture users as objectionable. Pastes and powders can be difficult to use because they require the denture wearer to ensure that the paste or powder is uniformly applied and in the right amount.

One approach for solving the problems of conventional denture adhesives is to provide a denture liner with mucoadhesive properties (10).

A denture liner with a hydrophobic polymer component, such as poly(ethyl methacrylate), an esterified copolymer of methyl vinyl ether and maleic anhydride, poly(vinyl acetate) has been developed (11). This type of liner inhibits the liner from absorbing moisture, by using a plasticizing component such as glycerol triacetate, or an adhesive component such as sodium carboxymethyl cellulose or poly(ethylene oxide). Carboxymethyl cellulose is shown in Figure 3.5.

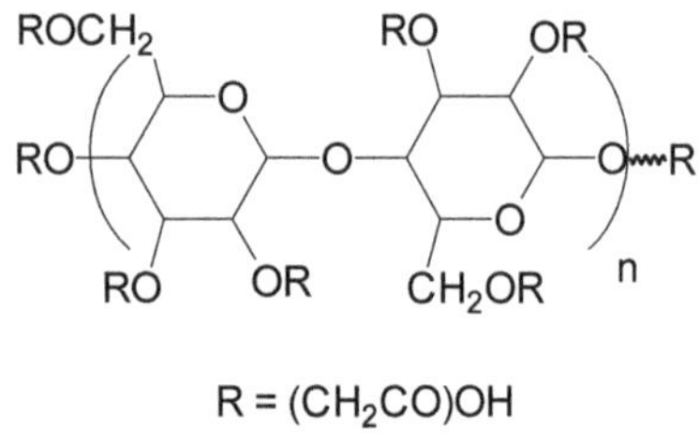

Figure 3.5 Carboxymethyl cellulose.

3.2.4.1 *Denture Adhesive Compositions*

Ordinary removable dentures, dental plates, and the like, are comprised of teeth mounted in a suitable plate or base. Denture stabilizers, including denture adhesives, are used to fill the interstices between the dentures and the gums or tissues. Prior to the placement of the denture in the oral cavity, a denture stabilizer is applied to the denture-plate surface, which, for a perfect fit, should uniformly contact the gums and mucous tissues. The denture stabilizer is formulated not only for its adherent properties, but also to provide a cushion or gasket between the denture and the gums or tissues, thereby positioning the denture securely in the oral cavity.

Denture adhesive compositions having good hold and an improved taste have been described that contain a salt of a copolymer of alkyl vinyl ether maleic acid (AVE/MA) or anhydride (12).

The alkyl vinyl ether maleic anhydride copolymers (AVE/MA anhydride) are obtained by the copolymerization of an alkyl vinyl ether monomer, such as methyl vinyl ether, ethyl vinyl ether, divinyl ether, propyl vinyl ether and isobutyl vinyl ether, with maleic anhydride to yield the corresponding alkyl vinyl ether maleic anhydride copolymer. This copolymer is readily hydrolyzable to the acid copolymer. When the anhydride copolymer dissolves in water, the anhydride linkage is cleaved so that the highly polar, corresponding polymeric free acid is formed. Accordingly, the anhydride form, which is relatively less expensive than the acid form, may be used as a convenient and cheaper precursor for the acid.

In addition, the salt form of the acidic polymers may be prepared by the interaction of the AVE/MA anhydride or acid copolymer with at least one cationic salt function, such as magnesium, strontium, or calcium, and optionally sodium. Also, compounds with a functional group, such as, hydroxide, oxide, acetate, halide, lactate, in an aqueous medium.

Both anhydride and acid forms are available from commercial suppliers. For example, the GAF Corporation, Parsippany, NJ., provides both the polymeric free acid form and the corresponding anhydride form under Gantrez™ (13).

Examples of compositions for salts of AVE/MA copolymers are shown in Table 3.4.

Table 3.4 Compositions for salts of AVE/MA copolymers (12).

Example	A	B	C	D
Component		Amount/[g]		
Water	1876.38	1883.68	1888.88	1889.58
Calcium hydroxide		11.39		7.59
Strontium hydroxide	43.62	24.93	24.93	18.7
Magnesium oxide			6.20	4.13
AVE/MA anhydride	80.00	80.00	80.00	80.00
Example	K	L	P	Q
			Comparative	
Component		Amount/[g]		
Water	1894.43	1896.41	1890.83	1892.77
Calcium hydroxide	23.95	25.83	25.28	27.12
Sodium hydroxide	3.98		7.80	27.12
AVE/MA anhydride	77.65	77.76	76.09	76.20

The compositions K and L in Table 3.4 exemplify salts of AVE/MA copolymers useful in the denture adhesive compositions. The comparative examples P and Q exemplify salts of AVE/MA copolymers having lower free acid levels that are used for a comparative check of the properties (12).

For a final composition, carboxymethyl cellulose was added to the salt composition of the AVE/MA copolymers. Also, an artificial saliva composition was prepared by adding several inorganic salts.

The compositions P and Q showed low cohesion, low tackiness, low elasticity, low re-healing, and high mealiness. The compositions K and L showed better properties.

Denture adhesive compositions in cream form can be fabricated by blending together the ingredients shown in Table 3.5.

Table 3.5 Denture adhesive compositions in cream form (12).

Ingredient	Weight/[*g*]
White Mineral Oil	89.74
Petrolatum, White	82.01
Carboxymethyl cellulose	75.00
Colloidal sodium silicon dioxide	4.28
Colorant (Opatint Red Dye)	0.23
An AVE/MA copolymer salt A-E, K,L	123.75

Also, formulations for other denture stabilizing compositions have been presented. These further include compositions in (12):

- Wafer form,
- Powder form, and
- Article or strip form.

3.2.4.2 *Hydrogel with Dry Tack*

In the past, poly(*N*-vinyl-2-pyrrolidone) (PVP) was only known as a processing aid for the manufacture of denture adhesives (10). It has been discovered that when PVP is combined with the cationic salts of a Grantrez hydrogel, a synergistic effect of promoting the dry tack of the hydrogel is created.

As a result of the synergistic combination of PVP and the cationic salts, the hydrogel not only provides improved adhesive and cohesive properties over an extended time, but also has a dry tack property (10). The term dry tack means that a denture adhesive with a good dry tack property provides an immediate bonding to surfaces.

Preferably, the monovalent cation is sodium and the bivalent cation is calcium. Also, a humectant, such as glycerine, is added. The humectant is used to provide a moistening effect to the hydrogel. The humectant also provides stability by limiting or controlling the rate at which the hydrogel denture adhesive absorbs water over time when placed in the mouth and thus controls the rate at which the mixed salts of AVE/MA copolymer hydrate and dissolve in the mouth.

Additional ingredients, which can be used in the denture adhesive, include therapeutically active agents suitable for treating individuals in need thereof. These agents are collected in Table 3.6.

Certain forms of therapy and combinations of these agents shown in Table 3.6 in the same delivery system may be useful in order to obtain an optimal effect.

Also, an anti-foaming agent, such as simethicone, can be included to eliminate the presence of unwanted gas bubbles without adversely affecting the adhesive properties of the hydrogel (10).

3.2.4.3 *Bioactive Dentures*

Chitosan-based bioactive containing acrylic materials has been reviewed (14). Bioactive materials with build-in capabilities for the treatment and the prevention of denture stomatitis have been described.

The surface of a denture base acrylic resin is porous, and denture plaque easily adheres to the surface of dentures (15). The colonization of microorganisms on the denture base acrylic resin occurs rapidly and *Candida* species strongly adheres to denture base materials.

Chitosan is a biologically safe biopolymer as well as an antioxidant. Chitosan has been proposed as a bioadhesive polymer. It is of continuous interest due to its unique properties and flexibility in a broad range of oral applications (16–18).

A bioactive gel was prepared by the dispersion of 0.2 *g* of commercially available bioactives (Propolis [Red Brazilian], Copaiba oil or Shiitake powder) in glycerol in an amount of 5% (1 *ml*) using a mortar and a pestle. Then 19 *ml* of glacial acetic acid (2%) was added with continuous mixing to the mixture of 0.2 *g* of commercially available bioactives (Propolis [Red Brazilian], Copaiba oil or Shiitake powder) in glycerol 5% (1 *ml*) and finally the chitosan polymer was added and mixed well to form the required gel and then mixed into a PMMA resin prior to setting. The amount of bioactive component, such as propolis, copaiba oil or shiitake mushroom extract, respectively, was determined to be 0.2 *g* of the prepared material (10 *g*) (14).

Table 3.6 Therapeutically active ingredients (10).

Antimicrobial agents	Antibiotics
Benzalkonium halides Quaternary ammonium salts Pyridinum salts Phosphonium salts Iodine Sulfonamides Bisbiguanides Phenolics	Tetracycline Neomycin Kanamycin Metronidazole Clindamycin
Anti-inflammatory agents	**Dental desensitizing agents**
Aspirin Acetaminophen Naproxen Ibuprofen Ketorolac Flurbiprofen Indomethacin Eugenol Hydrocortisone	Potassium nitrate Strontium chloride Sodium fluoride
Anesthetic agents	**Antifungals**
Lidocaine Benzocaine	Camphor Eucalyptus oil Benzaldehyde Insulin Steroids
Sweetening agents	**Coolants**
Menthol Menthyl lactate Wintergreen oil Peppermint oil Spearmint oil Leaf alcohol	3,1-Menthoxypropane-1,2-diol *N*-Ethyl-*p*-menthane-3-carboxamide

The reinforcement and bioactive addition of the copaiba oil, propolis (Brazilian) or shiitake mushroom extract of the tested PMMA dental resin resulted in no statistically significant increase of its compressive of tensile strength.

The additional benefits of the functionalized biomaterial as a preventative measure against biofilm formation, as well as build-in bioactive free radical defense capability of the materials, make them ideal candidates for further development and application in the materials for prosthetic devices (14).

3.3 Fabrication Methods

3.3.1 Rapid Prototyping in Dentistry

The term *rapid prototyping* was first used in the mechanical engineering field in the early 1980s to describe the act of producing a prototype, a unique product, the first product, or a reference model (19). In the past, prototypes were handmade by sculpting or casting, and their fabrication demanded a long time. Any and every prototype should undergo evaluation, correction of defects, and approval before the beginning of its mass or large-scale production. Prototypes may also be used for specific or restricted purposes, in which case they are usually called a preseries model.

With the development of information technology, three-dimensional models can be devised and built based on virtual prototypes. Computers can now be used to create accurately detailed projects that can be assessed from different perspectives in a process known as computer-aided design (19). In 1991, human anatomy models were produced with a technology called stereolithography (19).

Rapid mechanical prototyping can quickly fabricate complex shaped, 3D parts directly from computer-aided design models. The key idea of this technology is based upon decomposition of 3D computer models data into thin cross-sectional layers, followed by physically forming the layers and stacking them up.

This method has raised much attention in dentistry, especially in the field of surgery and implantology.

The historical development and various methods that are used for building dental appliances have been reviewed (20,21).

Also, the fundamentals and the developments of rapid prototyping and rapid manufacturing technologies and the application of the most common biomaterials, such as titanium and the titanium alloy (Ti_6Al_4V), have been discussed (22).

Besides the titanium and titanium alloys which were established as biocompatible materials, other biocompatible materials, such as cobalt-chromium and poly(ether ether ketone), have also been increasingly used in the fabrication of medical implants and dental prostheses.

As rapid prototyping technologies, selective laser sintering and selective laser melting are used along with fused deposition modeling. Also, electron beam melting could be successfully employed for the fabrication of medical implants and dental prostheses with complex features. In dentistry crown restoration, the use of thin copings of Ti_6Al_4V made by the electron beam melting process is an emerging trend (22).

The fabrication of medical implants and prostheses and biological models have three distinct characteristics (22,23):

1. Low volume,
2. Complex shapes, and
3. They are highly customized.

These characteristics make them suitable to be made by rapid manufacturing technologies, even on a commercial scale (22).

As in many branches of medicine, rapid prototyping has also been used in dentistry for a range of dental specialties (24), including oral and maxillofacial prosthodontics and surgery (25,26), dental implantology as a surgical guide or physical model and prosthodontics (23).

3.3.2 *Computer-Aided System*

The development of a computer-aided system for designing and fabricating removable complete dentures was investigated (27). The anatomical and morphological information of edentulous residual ridges and surrounding tissues were obtained from maxillo-mandibular double impressions mounted on a high-speed, three-dimensional laser scanner.

Artificial tooth arrangements, occlusion, the outline of polished surface, and denture border location were developed on a work station using a knowledge database. Established three-dimensional data were transferred into a three-dimensional laser lithography machine, and a complete denture was fabricated from a photopolymerized resin composite material (27).

The usage of computer-aided design/computer-aided manufacture, such as milling and rapid prototyping technologies for removable denture fabrication, has been reviewed (28). It can be concluded that the current innovations and technological developments allow the digital planning and manufacturing of removable dentures from start to finish (28).

The literature on computer-aided technology for fabricating complete dentures and which provides the reader with a historical background, current status, and future perspectives on this emerging technology has been detailed (29). An electronic search of the English language literature between the periods of January 1957 and June 2012 was performed.

A total of 1584 English language titles were found. Since the first published report in 1994, multiple authors have described different theoretical models and protocols for fabricating complete dentures with a computer-aided technology (29).

3.3.3 *Two-Step Impression for Complete Denture Fabrication*

The literature concerning the relevance of a two-step impression procedure to achieve better clinical results in fabricating conventional complete dentures has been reviewed (30).

Several aspects, such as oral health-related quality of life, patient satisfaction with dentures in use, masticatory performance and chewing ability, denture quality, direct and indirect costs, have been considered as eligible (30).

3.3.4 *Gingival Retraction Methods*

The methods of gingival retraction have been reviewed (31). The most common method used for gingival retraction was found to be chemomechanical. Actually, no method seemed to be significantly superior to the other in terms of gingival retraction achieved. The clinical parameters were not significantly affected by the gingival retraction method.

The successful fabrication of a fixed dental prosthesis requires that an appropriate impression be taken of the prepared finish line (32). This is critical in either tooth-supported fixed prosthesis (crown and bridge) or implant-supported fixed prosthesis (solid abutment).

If the prepared finish line is adjacent to the gingival sulcus, gingival retraction techniques should be used to decrease the marginal discrepancy between the restoration and the prepared abutment. Accurate marginal positioning of the restoration in the prepared finish line of the abutment is required for therapeutic, preventive and aesthetic purposes.

Gingival retraction techniques can be classified as mechanical, chemical or surgical. Conventional and recently developed methods of gingival retraction in the fixed tooth-supported prosthesis and fixed implant-supported prosthesis have been reviewed (32).

The application of astringent hemostatic agents is the most widely used technique for gingival retraction, and a variety of products are offered commercially. However, these products may have additional unintended yet clinically beneficial properties (33). The antimicrobial activities of marketed retraction products against plaque-associated bacteria in both planktonic and biofilm assays, have been assessed *in-vitro*. Hemostatic solutions, gels, pellets, retraction cords, and pastes have been tested.

All of the tested retraction products exhibited some antimicrobial activity. The results of the most active products were comparable to those of a marketed mouthwash. The listed retraction-active agents displayed relatively little activity when tested in pure form.

At 10% dilution, some products evidenced inhibitory activity against most tested bacteria within 3 *min* of exposure, whereas others displayed variable effects after 10 *min*. The most active agents reduced, but did not completely prevent, the metabolic activity of a monospecies biofilm (33).

References

1. A.S. Zafar-ul Islam and M. Fida, *Journal of the College of Physicians and Surgeons Pakistan*, Vol. 24, p. 791, 2014.

2. J.R. Elliot, G.M. Bowers, and B. Clemmer, *Journal of Periodontology*, Vol. 43, p. 217, 1972.
3. K.N. Rustogi, J.P. Curtis, A.R. Volpe, J.H. Kemp, J.J. McCool, and L.R. Korn, *The Journal of Clinical Dentistry*, Vol. 3, p. C9, 1992.
4. F. Ayad, S.A. Nathoo, A. Elias, B. Stewart, and F. Panakagos, Improved plaque identification index, US Patent Application 20 160 331 499, assigned to Colgate-Palmolive Company (New York, NY), November 17, 2016.
5. K. Ikeya, F. Iwasa, Y. Inoue, M. Fukunishi, N. Takahashi, K. Ishihara, and K. Baba, *The Journal of Prosthetic Dentistry*, Vol. in press, 2017.
6. K. Ikeya, M. Fukunishi, F. Iwasa, Y. Inoue, K. Ishihara, and K. Baba, *JoVE (Journal of Visualized Experiments)*, Vol. N.A., p. e54965, 2016.
7. K.E. Ahmed, K.Y. Li, and C.A. Murray, *Journal of Dentistry*, Vol. 61, p. 1, 2017.
8. S. Sadowsky and J.W. Stansbury, Denture tooth and material, US Patent 9 682 018, June 20, 2017.
9. A. Jokstad, M. Gökçe, and C. Hjortsjö, *International Journal of Prosthodontics*, Vol. 18, p. 489, 2005.
10. M.J. Borja, Denture adhesive hydrogel with dry tack, US Patent 9 408 780, assigned to Combe Incorporated (White Plains, NY), August 9, 2016.
11. M.J. Borja, Denture liner, denture liner kit and method for making a denture liner, US Patent 7 312 256, assigned to Combe Incorporated (White Plains, NY), December 25, 2007.
12. J. Rajaiah, R.S. Leonard, R.E. Bras, and F. Medeiros, Denture adhesive compositions, US Patent 9 463 145, assigned to The Procter & Gamble Company (Cincinnati, OH), October 11, 2016.
13. R.K. Haldar, B. Gangadharan, and R.K. Chaudhuri, Adhesive composition, US Patent 4 910 247, assigned to GAF Chemicals Corporation (Wayne, NJ), March 20, 1990.
14. T. Perchyonok, J. Souza, S. Zhang, D. Moodley, and S. Grobler, *International Journal of Medical Nano Research*, Vol. 2, p. 1, 11 2015.
15. M.A. Pfaller and D. Diekema, *Clinical Microbiology Reviews*, Vol. 20, p. 133, 2007.
16. S.R. Lockhart, S. Joly, K. Vargas, J. Swails-Wenger, L. Enger, and D.R. Soll, *Journal of Dental Research*, Vol. 78, p. 857, 1999.
17. V.T. Perchyonok, S. Zhang, and T. Oberholzer, *Current Organic Chemistry*, Vol. 16, p. 2437, 2012.
18. V. Tamara Perchyonok, S. Zhang, and T. Oberholzer, *Current Nanoscience*, Vol. 8, p. 541, 2012.
19. S. Nayar, S. Bhuminathan, and W.M. Bhat, *Journal of Pharmacy & Bioallied Sciences*, Vol. 7, p. S216, 2015.
20. Q. Liu, M.C. Leu, and S.M. Schmitt, *The International Journal of Advanced Manufacturing Technology*, Vol. 29, p. 317, Jun 2006.

21. A. Azari and S. Nikzad, *Rapid Prototyping Journal*, Vol. 15, p. 216, 2009.
22. W.P. Syam, M.A. Mannan, and A.M. Al-Ahmari, *Virtual and Physical Prototyping*, Vol. 6, p. 79, 2011.
23. R. Daule and V.N.V. Madhav, *Journal of Dental & Allied Sciences*, Vol. 2, p. 57, 2013.
24. H. Anderl, D.Z. Nedden, W. Mühlbauer, K. Twerdy, E. Zanon, K. Wicke, and R. Knapp, *British Journal of Plastic Surgery*, Vol. 47, p. 60, 1994.
25. S.M. Heckmann, M.G. Wichmann, W. Winter, M. Meyer, and H.-P. Weber, *Clinical Oral Implants Research*, Vol. 12, p. 617, 2001.
26. C.W. Hughes, K. Page, R. Bibb, J. Taylor, and P. Revington, *British Journal of Oral and Maxillofacial Surgery*, Vol. 41, p. 50, 2003.
27. Y. Maeda, M. Minoura, S. Tsutsumi, M. Okada, and T. Nokubi, *International Journal of Prosthodontics*, Vol. 7, p. 17, 1994.
28. M.S. Bilgin, E.N. Baytaroğlu, A. Erdem, and E. Dilber, *European Journal of Dentistry*, Vol. 10, p. 286, 2016.
29. A.S. Bidra, T.D. Taylor, and J.R. Agar, *The Journal of Prosthetic Dentistry*, Vol. 109, p. 361, 2013.
30. R.R. Regis, C.C.S. Alves, S.S.M. Rocha, W.A. Negreiros, and K.M. Freitas-Pontes, *Journal of Oral Rehabilitation*, Vol. 43, p. 771, 2016.
31. S. Tabassum, S. Adnan, and F.R. Khan, *Journal of Prosthodontics*, pp. n/a–n/a, 2016.
32. S. Safari, F.H. Ghavam, M. Hamedi, et al., *Journal of Dental Biomaterials*, Vol. 3, p. 205, 2016.
33. B. Hsu, S. Lee, D. Schwass, and G. Tompkins, *The Journal of the American Dental Association*, Vol. 148, p. 493, 2017.

4

Toothpaste Compositions

A toothpaste is a paste that is used with a toothbrush in order to clean and maintain the aesthetics and health of the teeth. A toothpaste serves as an abrasive that helps to remove dental plaque and food from the teeth.

Furthernore, it assists in suppressing halitosis, and delivers active ingredients, most commonly fluoride, to help prevent dental caries and gum disease, i.e., gingivitis (1). There are monographs dealing with the issues of toothpastes (2–4).

Teeth cleaning is an essential part of a typical daily hygiene routine and is a common practice (5). It is usually carried out with a toothpaste, consisting of abrasive particles in a carrier fluid, and a filament-based toothbrush, either manual or electric.

Oral care compositions generally refer to dentifrice and dental adhesives (6). Dentifrice formulations generally contain dentally acceptable abrasive, humectant, water, and water-soluble polymer which serves as a thickener and binder for the ingredients.

A variety of other ingredients, such as flavors, color, vitamins, antiplaque, anti-tarter, breath freshener, color, sweeteners, preservatives and fluoride, are also used at low levels.

Glycerol and sorbitol, cf. Figure 4.1, usually as an aqueous solution, are the most commonly used humectants for dentifrice, and depending on the characteristics desired in the product, poly(ethylene glycol) (PEG) or poly(propylene glycol) may be incorporated as well.

Four types of dentifrice are widely produced (6):

1. Cream,

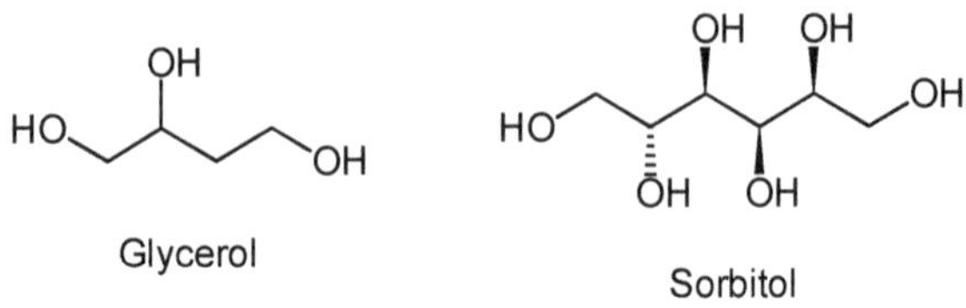

Figure 4.1 Humectants.

2. Transparent or translucent gel,
3. Stripes of cream and gel, and
4. Dry powders.

The thickeners or binders used for dentifrices are carboxymethyl cellulose (CMC), hydroxyethyl cellulose, silica, magnesium aluminum silicate, carrageenan, xanthan, guar, alginate, poly(acrylic acid)s (PAAs), salts of PAAs, polymers of ethylene oxide, copolymer of ethylene oxide and propylene oxide, processed euchemia seaweed, starch, starch derivatives, pectin, and agar (6).

In the manufacturing process for a dentifrice material, the incorporation of a dry water-soluble binder polymer into the composition often presents difficulties because of the tendency for lump formation when the dry polymers are added to and dispersed in aqueous systems. This increases the time required to obtain uniform hydration or uniform dispersion of the binder polymer. Sometimes, portions of the polymer could remain in dentifrice in a partially hydrated gel form.

This can not only affect dentifrice shelf stability and rheology but would also make dentifrice aesthetically unappealing. Another problem in the dentifrice industry is making a polymer that is universally dispersible, regardless of the environment in which it should be used (6).

4.1 History

Archaeologists have found a great deal of evidence that people have been practicing forms of oral hygiene for thousands of years. Chew sticks, tree twigs, bird feathers, and other implements have been used for teeth cleaning (5).

Already in 5000 BC, the Egyptians fabricated a tooth powder, which consisted of powdered ashes of ox hooves, myrrh, powdered and burnt eggshells, and pumice. Then, the Greeks, and afterwards the Romans, improved the recipes by adding abrasives such as crushed bones and oyster shells (1,7).

In the 9th century, the Iraqi musician and fashion designer Ziryab invented a type of toothpaste, which he popularized throughout Islamic Spain (8). The exact ingredients of this toothpaste are unknown, but it was reported to have been both functional and pleasant to taste.

It is not known whether these early toothpastes were used alone, were to be rubbed onto the teeth with rags, or were to be used with early toothbrushes, such as neem tree twigs and miswak (1). Toothpastes or powders came into general use in the 19th century. Fluoride was first added to toothpastes around 1890.

Glycerine was added early in the 19th century to make the powders into a paste and more palatable. Also, strontium was introduced at this time. It serves to strengthen teeth and reduce sensitivity. A dentist called Peabody became the first person to add soap to tooth powder in 1824 and chalk was added in the 1850s by John Harris (7).

The first mass-produced toothbrush is thought to have been made in England in 1750. Pig bristle was used for the cheap brushes, badger hair for the more expensive models. The first electric brush was invented in Switzerland in 1954 (5).

A list of toothpaste brands is collected in Table 4.1.

4.2 Ingredients for Toothpastes

4.2.1 *Abrasive Materials*

Toothpastes generally include an abrasive material which is dispersed in a gel or paste base. Abrasives remove stains and plaque, as well as polish teeth (11).

Common abrasives include calcium phosphates, alumina, calcium carbonate, and silica. Toothpaste must be abrasive enough to remove plaque and stains, but should not be so abrasive as to damage tooth enamel.

Table 4.1 List of toothpaste brands (9, 10).

Name	Year	Producer
Aim	1975	Unilever
Aquafresh	1973	GlaxoSmithKline
Arm & Hammer		
Babool	1987	Balsara Hygiene
Binaca		Dabur
Chlorodont	1949	
Close-Up	1967	Unilever
Colgate	1896	Colgate-Palmolive
Crest		Procter & Gamble product
Darlie	1933	Hawley & Hazel
Doramad	1940	Auergesellschaft of Berlin
Dr. Lyon's Tooth Powder	1920	Dr. J. W. Lyons D.D.S
Elmex	1962	GABA International AG
Euthymol		Johnson & Johnson
Gleem		Procter & Gamble
Ipana	1901	Bristol-Myers of New York
Kalodont		F. A. Sarg's Sohn & Co. Vienna
Macleans		GlaxoSmithKline
Marvis		
Mentadent		Unilever
Meswak	1998	Balsara Hygiene
Oral-B		Procter & Gamble
Patanjali		Baba Ramdev
Pepsodent		
Promise	1978	Balsara Hygiene
RA Thermoseal		ICPA Health Products, India
Sensodyne		GlaxoSmithKline
Signal		Unilever Company
Sozodont	1889	
Stomatol	1900	
Tom's of Maine	1970	Tom and Kate Chappell
Ultra Brite		Colgate-Palmolive
Unique		ACS Manufacturing Corp.
Zendium		Unilever

4.2.1.1 *Calcium-Based Abrasives*

Toothpastes are generally classified into two types (12): Opaque and transparent. Opaque toothpastes are usually white or colored. Such toothpastes usually contain calcium-based abrasives, in particular chalk. Gel toothpastes usually contain abrasive silica. Also known are multiphase toothpastes, which have an opaque phase and a transparent phase.

Toothpastes, in particular-chalk based types, also contain thickening silica. Silica helps build up the viscosity. However, the thickening silica-based agent, especially at higher levels, may adversely affect the flavor delivery. This often compels formulators to dose an appropriate overage of the flavor. Since flavors are very expensive, even a slight reduction in flavor can be economically significant. Therefore, it is desirable to either reduce the level of thickening silica, or preferably, have no thickening silica.

Toothpastes usually contain some polymers. Cellulosic polymers, such as sodium carboxymethyl cellulose, are included as thickeners. Polymers of vinyl methyl ether and maleic acid are often used in toothpastes that contain chalk. Such polymers are used for the delivery of flavor.

One of the disadvantages of polymers of vinyl methyl ether and maleic acid, especially in toothpastes containing calcium-based abrasives, is that the unbalanced amount of such polymers can affect the spreadability of the paste. Furthermore, the paste becomes unstable after about a month of storage at elevated temperature. An arbitrary reduction in the level of such polymers cannot be a viable solution, as it may adversely affect delivery of flavor.

It has been found that toothpastes containing a calcium-based abrasive, a copolymer of vinyl methyl ether and maleic acid and clay have significantly better stability even at elevated temperatures.

A preferred abrasive is fine ground natural chalk (FGNC), which is a form of chalk. Cellulosic polymers, such as sodium carboxymethyl cellulose (SCMC), are included as thickeners. MFIL® is a thickening silica compound. Veegum® belongs to the group of natural and synthetic clays. Some formulations of the toothpastes are described in Tables 4.2 and 4.3.

After making the toothpastes, they were stored at 45°C for one month, after which their stability was checked.

Table 4.2 Toothpaste formulations (12).

Composition No.:	1	2	3	4
Compound	[%] w/w	[%] w/w	[%] w/w	[%] w/w
Water	30	32	30	28
Veegum® HV	–	–	–	0.8
Sorbitol (70%)	15	15	15	15
FGNC	40	40	45	45
MFIL®	3.75	3.75	–	1.5
SCMC	0.625	0.625	0.9	0.45
Gantrez™ S-97	2	0.5	0.8	0.75
Sodium lauryl sulfate	2.5	2.5	2.5	2.5
Sodium silicate (30%)	1.75	1.75	1.75	1.75
Flavor	0.2	0.2	0.2	0.2

Table 4.3 Toothpaste formulations (12).

Composition No.:	5	6	7
Compound	[%] w/w	[%] w/w	[%] w/w
Water	25	34	34
Veegum® HV	0.8	0.8	0.8
Sorbitol (70%)	15	15	15
FGNC	50	40	40
MFIL®	–	1.5	1.5
SCMC	0.45	0.45	0.45
Gantrez™ S-97	0.75	0.3	0.75
Sodium lauryl sulfate	2.5	2.5	2.5
Sodium silicate (30%)	1.75	1.75	1.75
Flavor	0.2	0.2	0.2

Composition 1 with 2% Gantrez and 3.75% thickening silica was rubbery. This was believed to be because of the high level of Gantrez S-97. When Gantrez was reduced to 0.5% in composition 2, the composition was acceptable when made, but turned rubbery after storage. This indicates that reduction in Gantrez did not solve the technical problem. Composition 3 was made without thickening silica. However, this composition was also found to be unstable, and so were compositions 4 and 5. Only the preferred compositions 6 and 7 having selective balance between the calcium-based abrasive, the copolymer of vinyl methyl ether and maleic acid, and the clays were stable (12).

4.2.2 *Fluoride*

Fluoride is typically added to a toothpaste in order to reduce tooth decay (11). In particular, fluoride incorporates itself into tooth enamel to make teeth more resistant to acids produced by plaque bacteria, as well as acids found in fruit juices, soda and certain foods.

Actually, toothpastes containing fluoride hardens tooth enamel to make the entire tooth structure more resistant to decay and promote remineralization, which aids in repairing early decay. In toothpaste, fluoride is commonly found in the form of sodium monofluorophosphate, stannous fluoride, or sodium fluoride.

Due to the toxicity of fluoride, the Food and Drug Administration (FDA) regards any toothpaste that contains fluoride compound as a drug.

Accordingly, the FDA requires a warning on the label of any toothpaste containing fluoride which states: *If you accidentally swallow more than used for brushing, seek professional help or contact a poison control center immediately.*

Moreover, the American Dental Association (ADA) requires that toothpaste manufacturers include the following language on all ADA approved toothpastes containing fluoride: *Do not swallow. Use only a pea-sized amount for children under six. To prevent swallowing, children under six years of age should be supervised in the use of toothpaste.*

Accordingly, toothpastes containing fluoride are not intended to, and should not, be swallowed (11).

4.2.3 *Detergents*

Detergents may be added to toothpastes to aid the cleaning properties (11). Detergents may be added to create a foaming action. The foam prevents the toothpaste from dribbling out of the mouth during brushing. Sodium lauryl sulfate is a commonly used detergent. Sodium lauryl sulfate is also known as sodium dodecyl sulfate

4.2.4 *Humectants*

A humectant provides mouthfeel and also prevents the toothpaste composition from drying out (13). Typical humectants are polyols of three to six carbons in which each carbon is hydroxylated, and mixtures thereof, such as glycerol (glycerin), sorbitol, poly(ethylene glycol), poly(oxyethylene glycol), mannitol, xylitol, and other sugar alcohols. Sorbitol and glycerol are preferred. The water is preferably deionized and free of impurities.

4.2.5 *Thickeners*

Thickeners can prevent the growth of microorganisms. Gum base can be included to thicken the final toothpaste product so as to result in a paste, rather than a liquid. Gum base may include chicle, xanthan gum, guar gum, or a poly(butene).

Chicle is a natural gum that is traditionally used for the fabrication of chewing gum. It is collected from several species of Mesoamerican trees (14). Both the Aztecs and Maya traditionally chewed chicle. Chicle was chewed as a way to stave off hunger, freshen breath, and keep the teeth clean.

Also, a cellulose gum, such as carboxymethyl cellulose, or xanthan gum may be used (11). Such a gum base can be prepared by dissolving carboxymethyl cellulose in hot water, adding xanthan gum and mixing. Xanthan gum can be produced by the fermentation of glucose, sucrose, or lactose. Xanthan gum is shown in Figure 4.2.

4.2.6 *Sweeteners*

There is a wide variety of compounds that enhance the sweetness (15). Although naturally occurring carbohydrate sweeteners, such

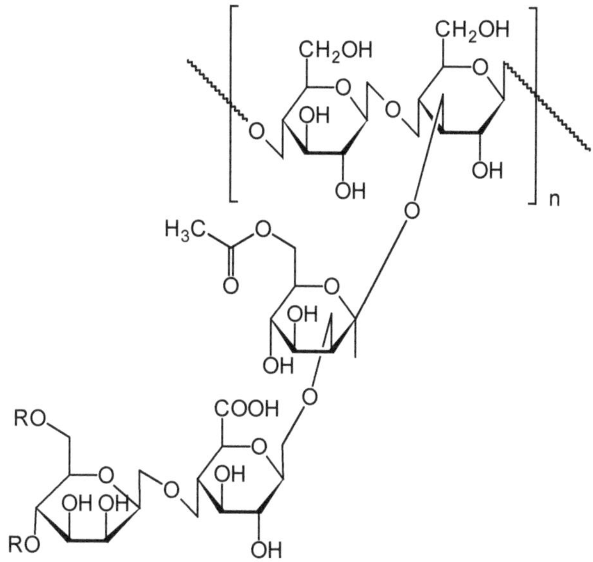

Figure 4.2 Xanthan gum.

as sucrose, are the most widely used sweeteners, they suffer from the disadvantages of high cost, high caloric content, and the promotion of tooth decay.

Artificial sweeteners have been designed that overcome these problems but they are sometimes rejected by the consumer for not having a sufficiently *sucrose-like* taste. Artificial sweeteners have different sweetness profiles from that of sucrose and often suffer from side effects such as delays in the onset of sweetness perception or unpleasant aftertastes (15). Some sweeteners are shown in Table 4.4 and in Figure 4.3.

High-intensity sweeteners are commonly used as sugar substitutes or sugar alternatives because they are much sweeter than sugar but contribute only a few to no calories when added to foods (16). The sweetnesses and energy densities of sugar substitutes have been listed in comparison to those of sucrose (17).

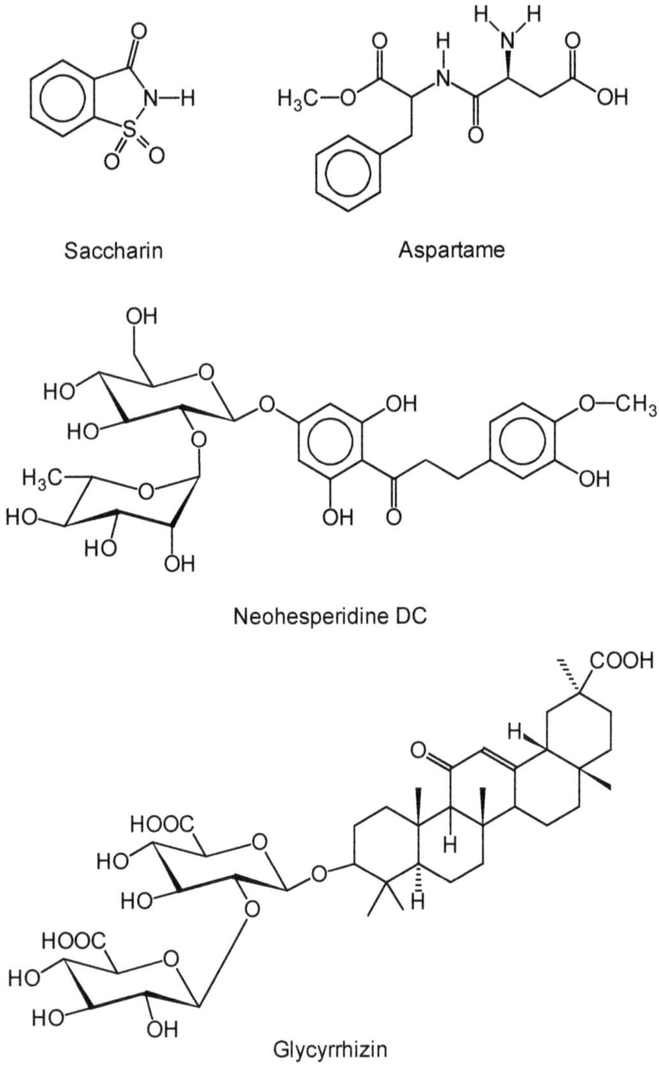

Figure 4.3 Sweeteners (18).

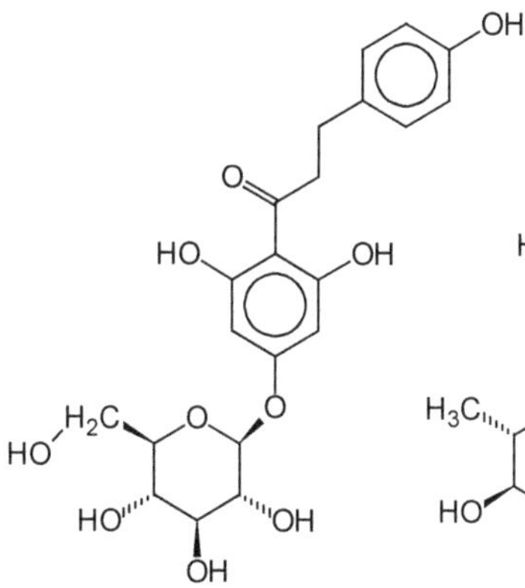

Trilobtain

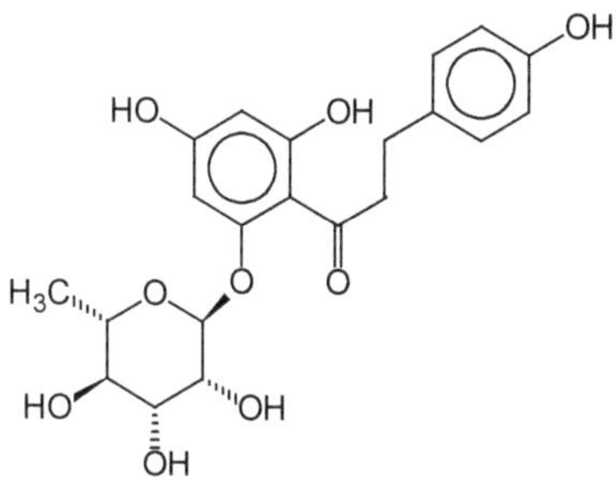

Glycyphyllin

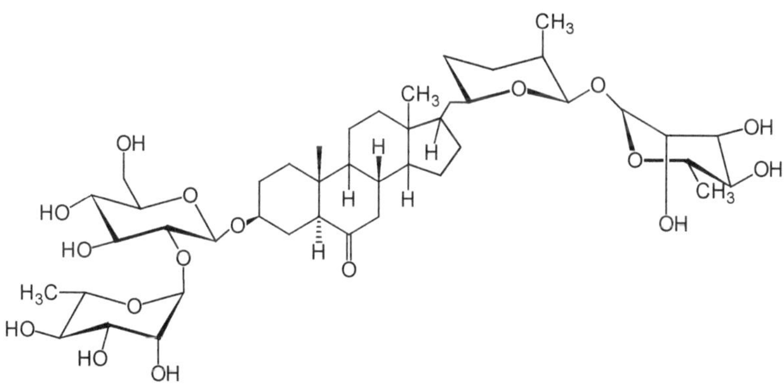

Osladin

Figure 4.3 (cont.) Sweeteners (18)

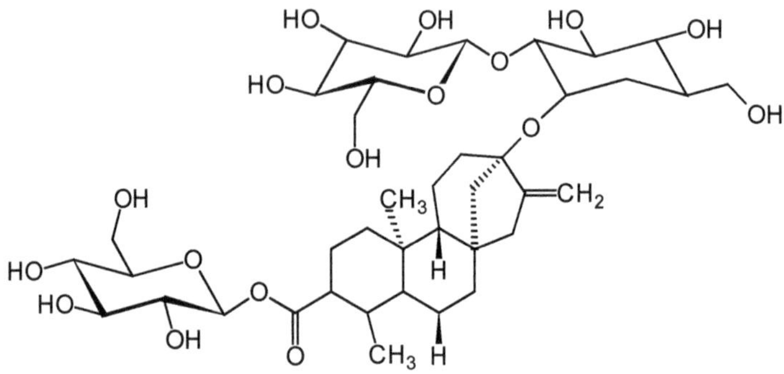

Stevioside

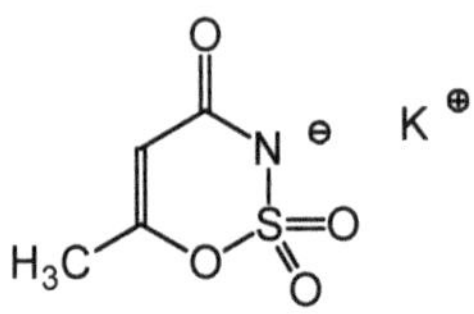

Acesulfame K

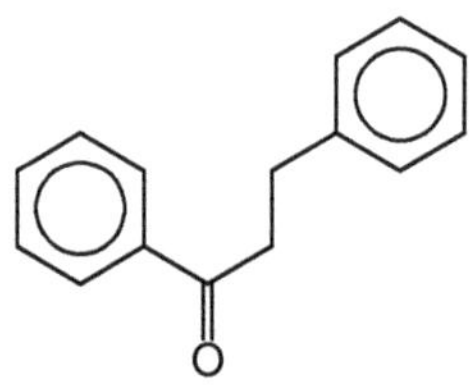

Dihydrochalcone

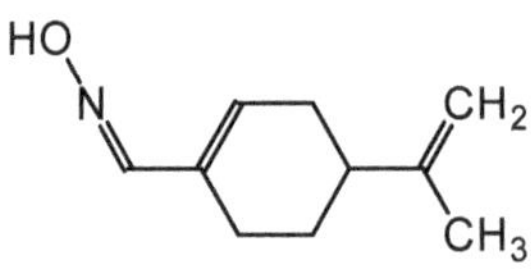

Perillartine

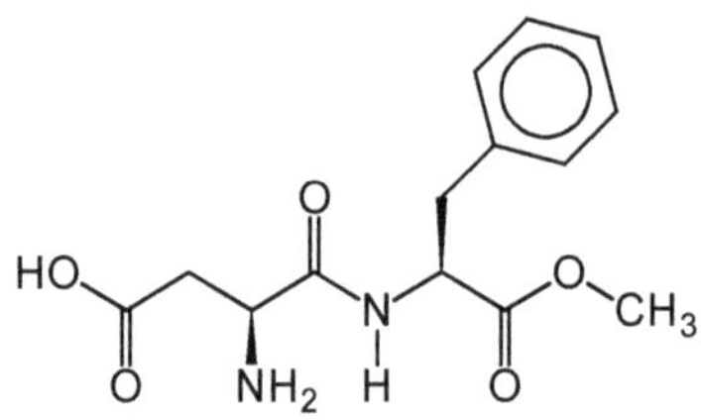

Aspartylphenylalanine methyl ester

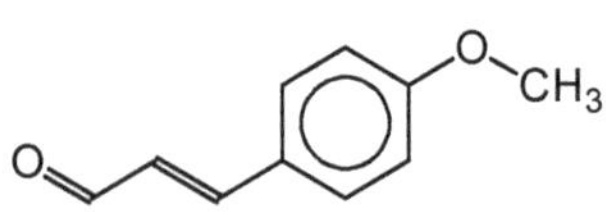

4-Methoxycinnamic aldehyde

Figure 4.3 (cont.) Sweeteners (18)

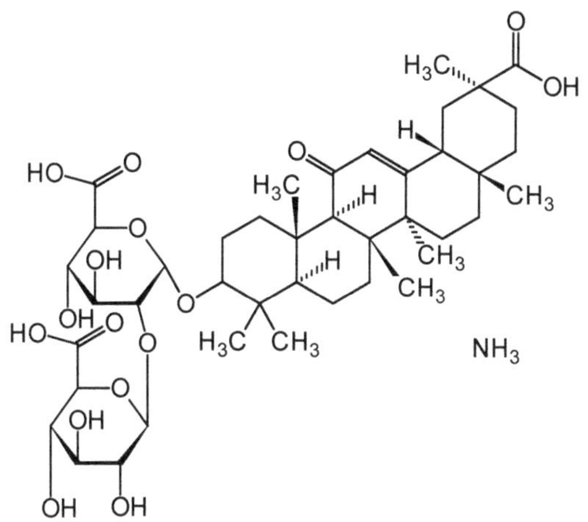

Glycyrrhizin

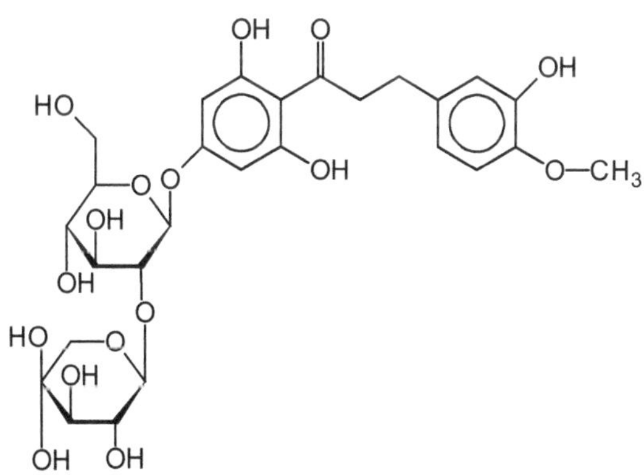

Neo-hesperidyl dihydrochalcone

Figure 4.3 (cont.) Sweeteners (18)

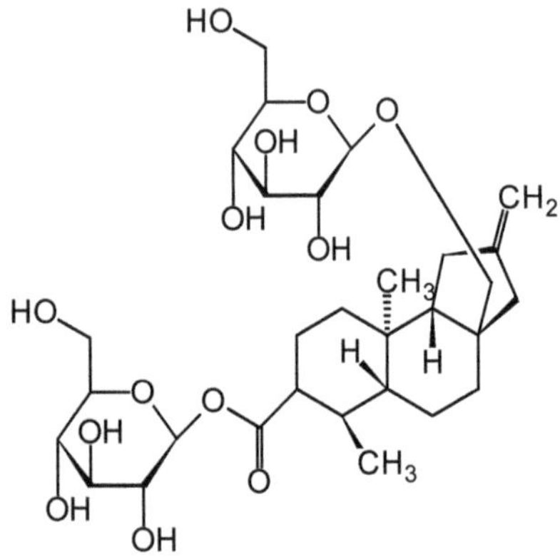

Rubusoside

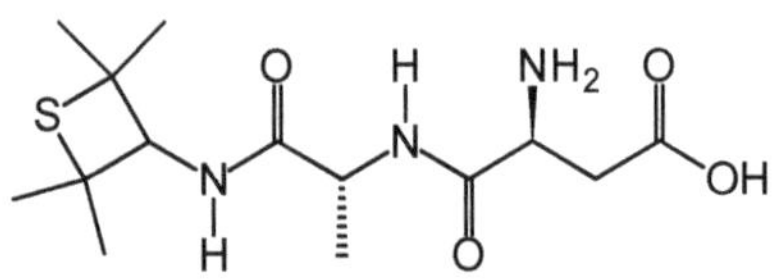

Alitame

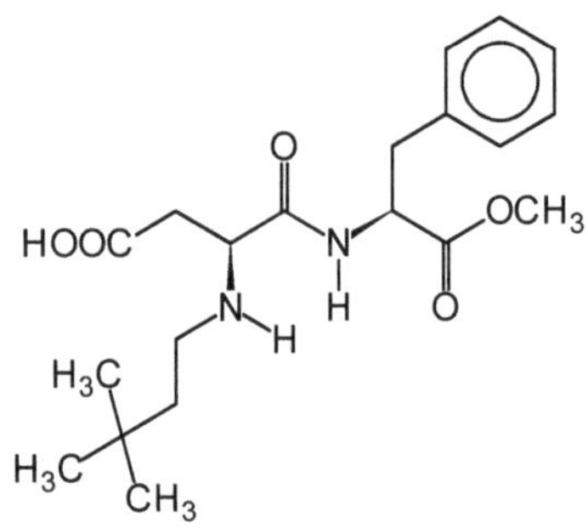

Neotame

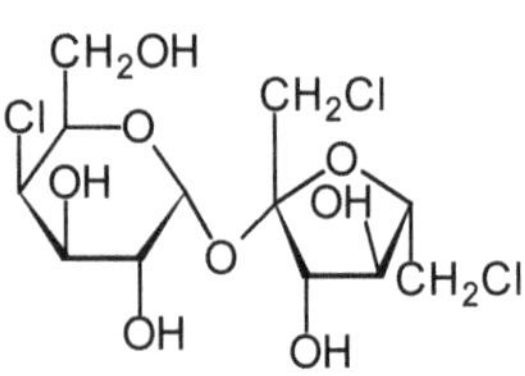

Sucralose

Figure 4.3 (cont.) Sweeteners (15)

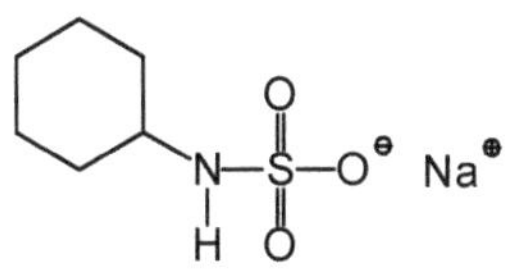

Cyclamate

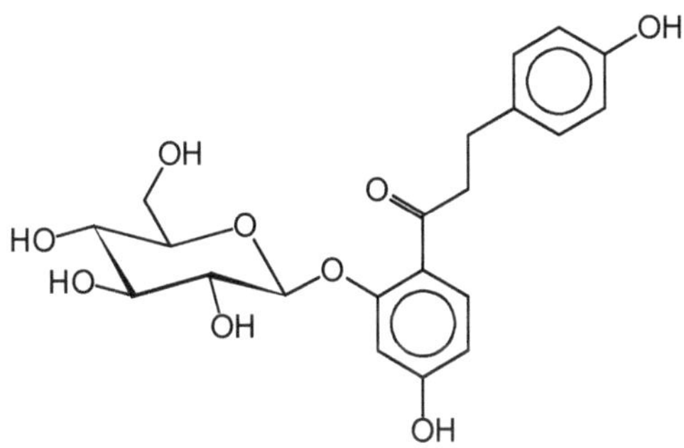

Phloridzin

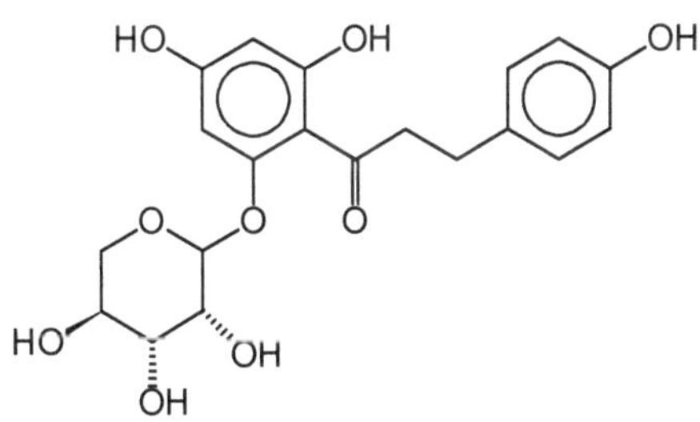

Glycyphyllin

Figure 4.3 (cont.) Sweeteners (15)

Table 4.4 Sweeteners (15).

Compound	Compound	Compound
Saccharin	Alitame	Aspartame
Neotame	Cyclamate	Sucralose
Acesulfame	Neohesperidine DC	Thaumatin
Glycyrrhizin	Mogroside IV	Mogroside V
Cyclocarioside I	Rebaudiosides	Dulcosides
Rubusoside	Stevia	Stevioside
Luo Han Guo	Siamenoside	Monatin
Curculin	Monellin	Mabinlin
Brazzein	Hemandulcin	Phyllodulcin
Glycyphyllin	Phloridzin	Trilobtain
Baiyanoside	Osladin	Polypodoside A
Pterocaryosides	Mukurozioside	Phlomisoside I
Periandrin I	Abrusoside A	

4.2.7 *Coloring Agents*

Colored products are usually formulated by merely adding the desired dye to the other components prior to the mixing stage of the process (19). As the dyes utilized in toothpaste are all water-soluble, and varying large amounts of water are present in toothpaste, the desired color spreads and uniformly colors the entire product.

An alternative means of uniformly coloring a toothpaste is to uniformly disperse in the paste very small particles of insoluble colored material which acts as a pigment. Although this material is discerned as discrete colored particles when the paste is examined under a magnifying lens, to the naked eye the paste has a uniformly colored appearance. For example, colored thermosetting crosslinked resin particles can be used for uniform pigmentation of toothpastes (19). Some coloring agents are summarized in Table 4.5 and are also shown in Figure 4.4.

4.2.8 *Organic Antimicrobial Agents*

Examples of organic antimicrobial agents are quaternary ammonium compounds (21). Such are those in which one or two of the substitutes on the quaternary nitrogen have a carbon chain length from about 10 to about 18 carbon atoms while the remaining substi-

Table 4.5 Coloring agents (20).

Compound	Synonym
Yellow No. 4	Tartrazine
Yellow No. 201	Fluorescein
Yellow No. 202(1)	Uranine
Yellow No. 202(2)	Uranine K
Yellow No. 203	Quinoline Yellow WS
Yellow No. 204	Quinoline Yellow SS
Yellow No. 205	Benzidine Yellow G
Yellow No. 401	Hanza Yellow
Yellow No. 402	Pola Yellow 5G
Yellow No. 403(1)	Naphthol Yellow S
Yellow No. 406	Metanyl Yellow
Yellow No. 407	Fast Light Yellow 3G
Hansa Yellow 10G	
Disazo Yellow	AAMX, AAOT, HR, 4G, 3A, GR, G
Benzimidazolone Yellow	H2G, HG
Isoindoline Yellow	G, R
Pyrazolone Yellow HGR	
Diarylide Yellow AAOA	
Green No. 3	Fast Green FCF
Blue No. 1	Brilliant Blue FCF
Blue No. 2	Indigo Carmine
Blue No. 201	Indigo
Blue No. 202	Patent Blue NA
Bluc No. 203	Patent Blue CA
Blue No. 204	Carbanthrene Blue
Blue No. 205	Alphazurine FG
Phthalocyanine Blue	
Aluminum Phthalocyanine Blue	
Indanthrene Blue	

Table 4.5 (cont.) Coloring agents (20)

Compound	Synonym
Red No. 2	Amaranth
Red No. 104	Phloxine
Red No. 105	Rose Bengal
Red No. 106	Acid Red
Red No. 201	Lithol Rubine B
Red No. 202	Lithol Rubine BCA
Red No. 203	Lake Red C
Red No. 204	Lake Red CBA
Red No. 205	Lithol Red
Red No. 206	Lithol Red CA
Red No. 207	Lithol Red BA
Red No. 208	Lithol Red SR
Red No. 213	Rhodamine B
Red No. 214	Rhodamine B Acetate
Red No. 215	Rhodamine B Stearate
Red No. 218	Tetrachlorotetrabromofluorescein
Red No. 219	Brilliant Lake Red R
Red No. 220	Deep Maroon
Red No. 221	Toluidine Red
Red No. 223	Tetrabromofluorescein
Red No. 225	Sudan III
Red No. 226	Helindone Pink CN
Red No. 227	Fast Acid Magenta
Red No. 228	Permaton Red
Red No. 230(1)	Eosin YS
Red No. 230(2)	Eosin YSK
Red No. 231	Phloxine BK
Red No. 232	Rose Bengal K
Red No. 401	Violamine R
Red No. 404	Brilliant Fast Scarlet
Red No. 405	Permanent Red F5R
Red No. 501	Medical Scarlet
Red No. 502	Ponceau 3R
Red No. 503	Ponceau R
Red No. 504	Ponceau SX
Red No. 505	Oil Red XO
Red No. 506	Fast Red S
Purple No. 201	Alizurine Purple Lake SS
Purple No. 401	Alizurol Purple
Naphthol AS	

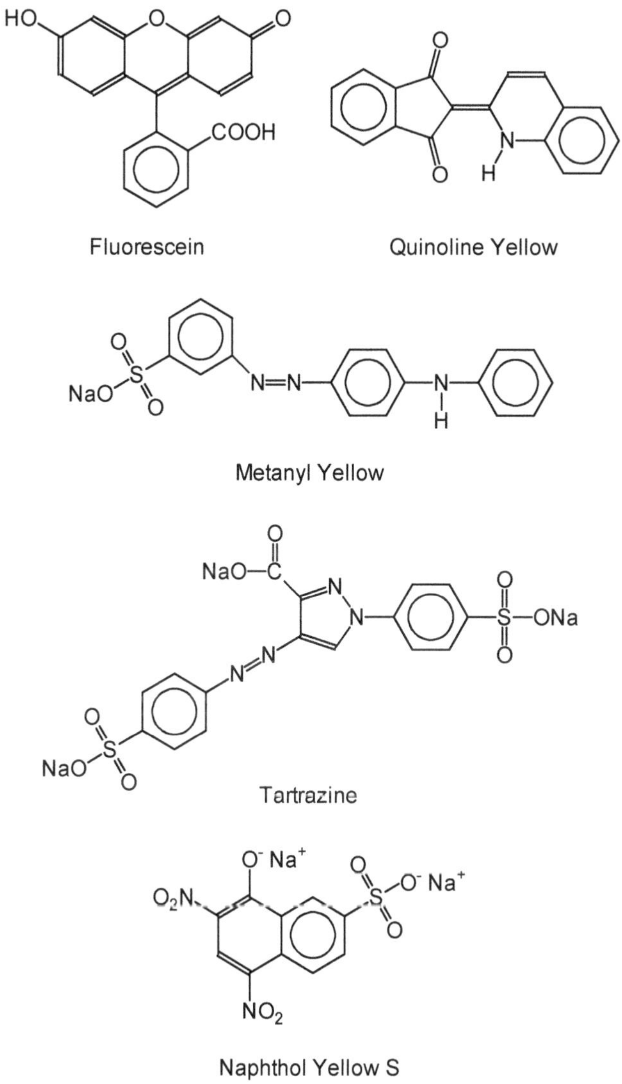

Figure 4.4 Coloring agents.

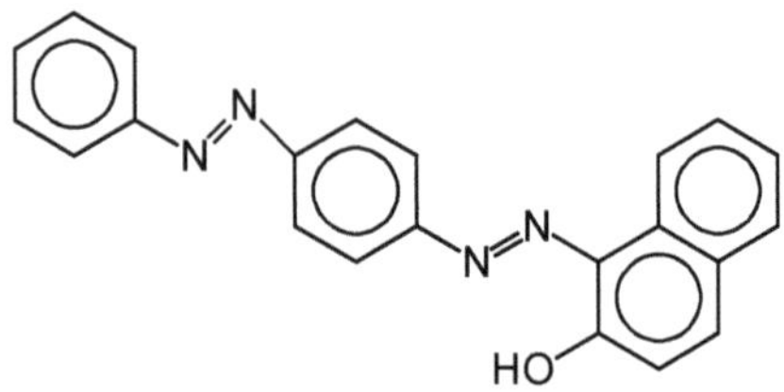

Sudan III

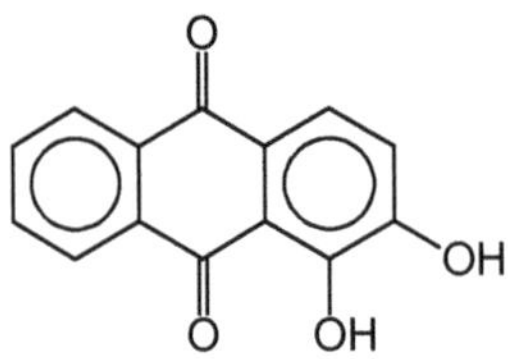

Alizarin

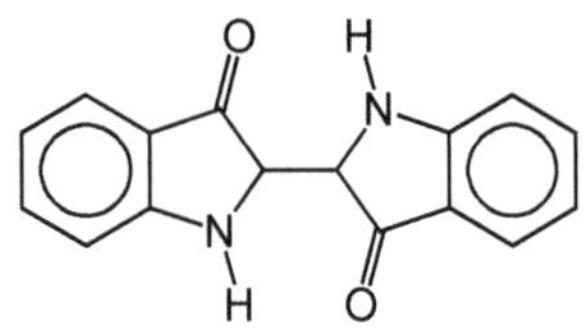

Indigo

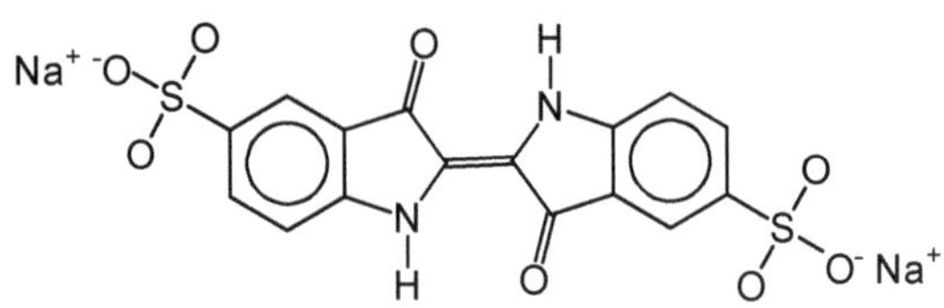

Indigo Carmine

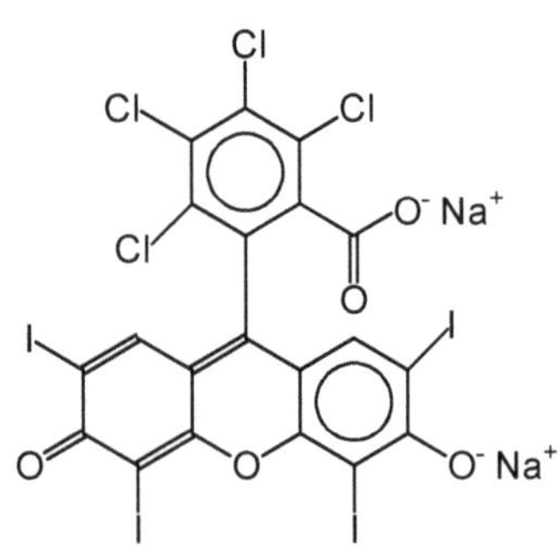

Rose Bengal

Figure 4.4 (cont.) Coloring agents

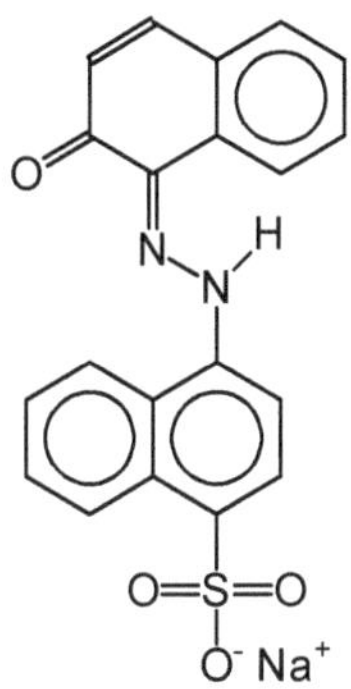

Fast Red S

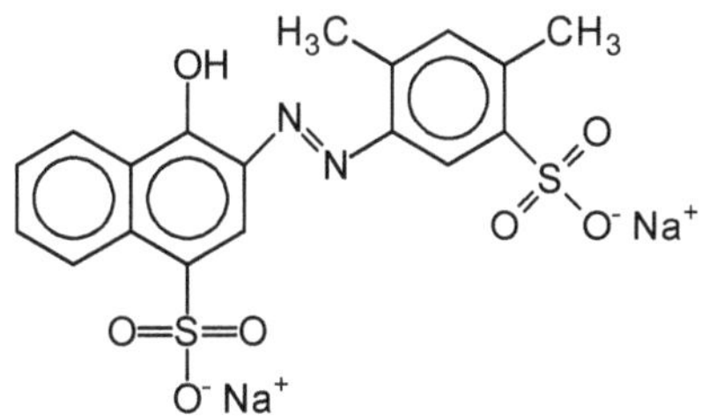

Ponceau SX

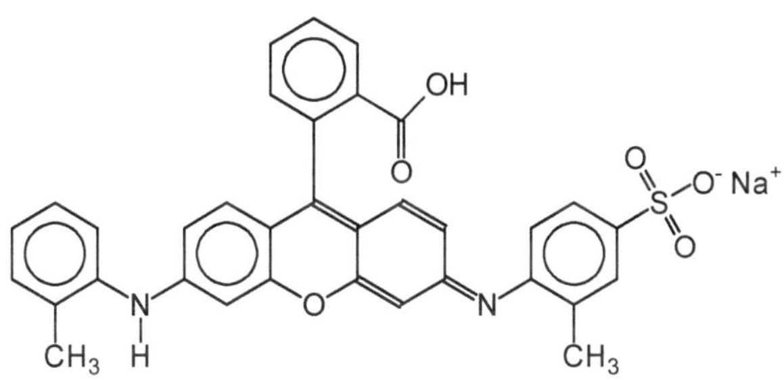

Violamine R

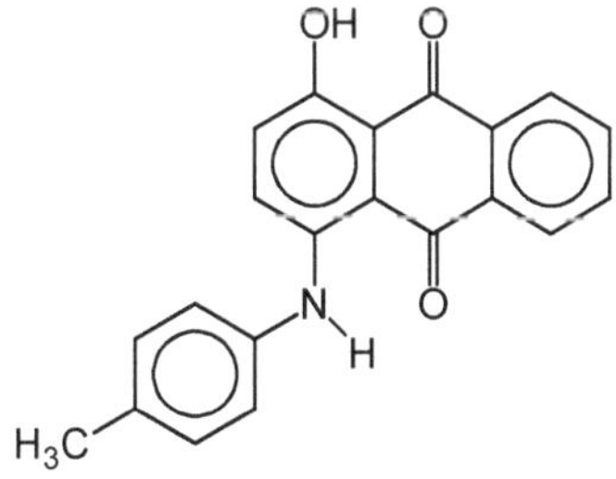

Alizurol purple

Figure 4.4 (cont.) Coloring agents

tutes, such as alkyl or benzyl groups, have a lower number of carbon atoms, typically methyl or ethyl groups. Examples are collected in Table 4.6 and in Figure 4.5.

Table 4.6 Organic antimicrobial agents (21).

Compound
Dodecyl trimethylammonium bromide
Tetradecylpyridinium chloride
Domiphen bromide
N-Tetradecyl-4-ethyl pyridinium chloride
Dodecyl dimethyl(2-phenoxyethyl) ammonium bromide
Benzyl dimethoylstearyl ammonium chloride
Cetylpyridinium chloride
Quaternized 5-amino-1,3-bis(2-ethyl-hexyl)-5-methyl hexahydropyrimidine
Benzalkonium chloride
Benzethonium chloride

Other useful organic antimicrobials are enzymes such as endoglycosidase, papain, dextranase, mutanase, and combinations thereof (21).

4.2.9 *Bad Breath Reduction Agents*

Bad breath reduction agents are used to reduce breath malodor (21). Examples of these agents are shown in Table 4.7.

4.2.10 *Flavoring Agents*

Examples of flavoring agents are shown in Table 4.8.

4.3 Compositions

4.3.1 *Processes for Preparing Toothpaste Compositions*

Actually, there are problems associated with filling a toothpaste package effectively and reliably if the toothpaste exhibits the stringy effect and tailing, in particular for toothpaste formulations containing a significant proportion of calcium carbonate particles (22).

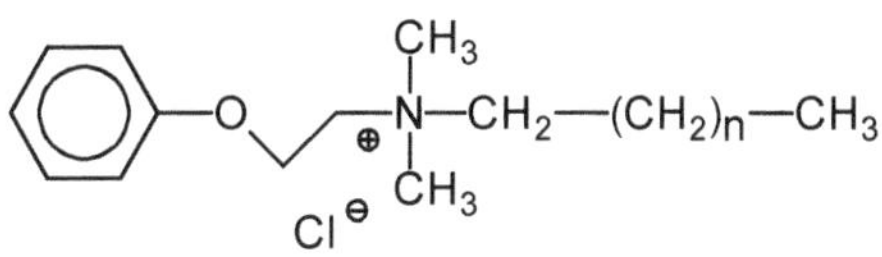

Benzalkonium chloride

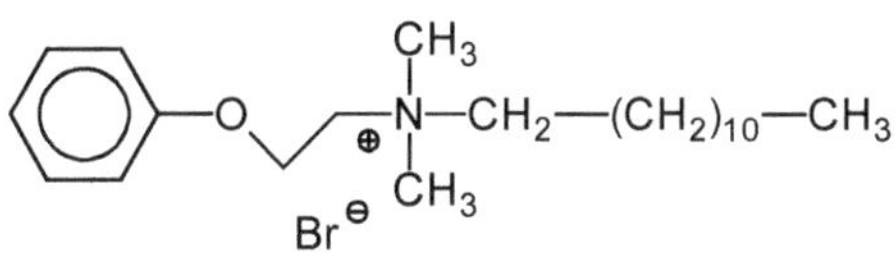

Domiphen bromide

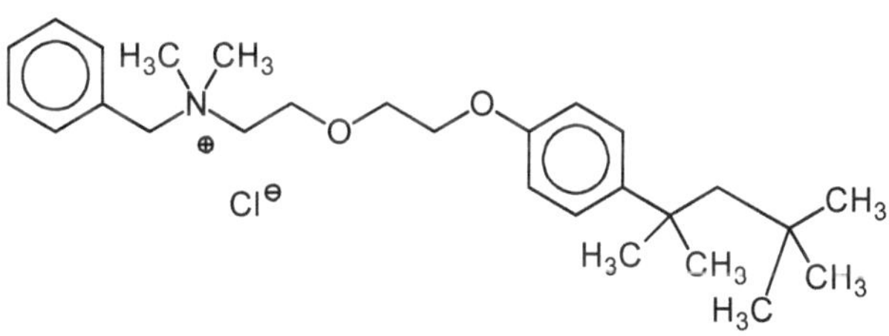

Benzethonium chloride

Figure 4.5 Organic antimicrobial agents.

Table 4.7 Bad breath reduction agents (21).

Compound
Ascorbic acid [3-oxo-L-gulofuranolactone]
cis-Jasmone[3-methyl-2-(2-pentenyl)-2-cyclopentenone]
2,5-Dimethyl-4-hydroxy-3(2H)-furanone
5-Ethyl-3-hydroxy-4-methyl-2(5H)-furanone
Vanillin[3-hydroxy-3-methoxybenzaldehyde]
Ethyl vanillin
Anisaldehyde[4-methoxybenzaldehyde]
3,4-Methylenedioxybenzaldehyde
3,4-Dimethoxybenzaldehyde
4-Hydroxybenzaldehyde
2-Methoxybenzaldehyde
Benzaldehyde
Cinnamaldehyde[3-phenyl-2-propenal]
Hexyl cinnamaldehyde
α-Methyl cinnamaldehyde
o-Methoxy cinnamaldehyde
Citral
Linalool
Geraniol
Eugenol

Table 4.8 Flavoring agents (21).

Compound	Compound
Mint oils	Wintergreen
Clove bud oil	Cassia
Sage	Parsley oil
Marjoram	Lemon
Orange	Propenyl guaethol
Heliotropine	4-*cis*-Heptenal
Diacetyl	Methyl salicylate
Ethyl salicylate	1-Menthyl acetate
Oxanone	α-Irisone
Methyl cinnamate	Ethyl cinnamate
Butyl cinnamate	Ethyl butyrate
Ethyl acetate	Methyl anthranilate
i-Amyl acetate	*i*-Amyl butyrate
Allyl caproate	Eugenol
Eucalyptol	Thymol
Cinnamic alcohol	Octanol
Octanal	Decanol
Decanal	Phenylethyl alcohol
Benzyl alcohol	α-Terpineol
Linalool	Limonene
Citral	Neral
Geranial	Geraniol nerol
Maltol	Ethyl maltol
Anethole	Dihydroanethole
Carvone	Menthone
β-Damascenone	Ionone
γ-Decalactone	γ-Nonalactone

The traditional manufacturing processes result in toothpastes having a granular texture, which is not satisfactory because the consumers require toothpaste compositions which are smooth in texture.

Therefore, a method is needed for manufacturing a toothpaste composition which contains calcium carbonate particles and also exhibits good rheological properties, in particular one which reduces stringiness and tailing and has a smooth texture.

In addition, there is a well-known way to manufacture striped toothpaste, in which a first toothpaste phase is provided with stripes of a second phase, of toothpaste or gel, when the toothpaste is extruded as a strip from a tube or other packaging. The first toothpaste phase is typically white and the second phase is typically colored.

It is an important aesthetic requirement of the consumer that the different colored phases have a sharp boundary between them and that the different colorants do not appear to bleed from one phase into the adjacent phase.

The binder component in the toothpaste has a significant impact on the stripe quality of the extruded strip to form a stable extrudable paste including the abrasive particles and the liquid phase. In striped toothpaste formulations containing a significant proportion of calcium carbonate particles, it is known to use a binder system which controls the stripe quality. In particular, it is known to use a binder system incorporating the combination of CMC and magnesium aluminium silicate (MAS) to provide a high stripe quality. However, the use of MAS suffers from the problem that currently there are only a few qualified suppliers of toothpaste-grade MAS, which increases the cost and complexity of the supply chain.

A method has been described for manufacturing a toothpaste composition which contains calcium carbonate particles and exhibits good rheological properties, in particular reduced stringiness and reduced tailing, and has a smooth texture. The method consists of the following steps (22):

1. Dispersing guar gum into a first aqueous medium comprising sorbitol to produce a first gellant dispersion including hydrated guar gum in a sorbitol solution,

2. dispersing at least one cellulose polymer into a second aqueous medium to produce a second gellant dispersion including hydrated cellulose polymer in an aqueous solution,
3. Combining the first and second gellant dispersions to form a third gellant dispersion, and
4. Mixing the third gellant dispersion with a plurality of toothpaste components, the toothpaste components including an abrasive comprising calcium carbonate particles, to form a toothpaste composition.

In a preferred composition, the guar gum is present in an amount of about 0.15% based on the weight of the toothpaste composition and CMC is present in an amount of about 1% based on the weight of the toothpaste composition. Sorbitol may be present in an amount of 12–18% based on the weight of the toothpaste composition.

Exemplary compositions are shown in Table 4.9.

The toothpaste composition produced according to the method of Example 1 exhibited a smooth non-granular texture.

The viscosity of the composition of Example 1 was evaluated in an aging test over a period of 1 week. The initial viscosity and final viscosity, after a period of 1 week, of the composition of Example 1 were acceptable for use as a commercial toothpaste.

Furthermore, the rheological properties of composition of Example 1 were found to provide a toothpaste composition which stays on the mixer blade as a common non-flowing mass.

In contrast, some toothpaste compositions incorporating calcium carbonate and other binder systems than guar gum/CMC exhibit a *tailing* phenomenon in which the toothpaste composition flows under gravity to form trails extending downwards from the blade. The resultant toothpaste can be difficult to use in a package filling process. Accordingly, it may be seen that the addition of guar gum to the CMC binder composition made also solves the problem of trailing of the composition. The compositions so produced solve the trailing issue in toothpaste compositions including calcium carbonate and CMC.

Other rheological properties of the composition were tested, in particular the flow property, thixotropy, yield stress and creep recovery, and these results showed that the binder system of CMC and

Table 4.9 Toothpaste composition (22).

Ingredient Example 1	[%] w/w
Sorbitol (70% aqueous solution)	21
Guar gum	0.15
Sodium saccharin	0.27
Sodium monofluorophosphate	1.1
CMC	1
Sodium bicarbonate	0.1
Sodium carbonate	0.4
Silica	2
Calcium carbonate	42
Titanium dioxide	0.1
Sodium lauryl sulfate	2
Methylparaben	0.1
Propylparaben	0.02
Water	28.76
Flavor	1
Ingredient Colored Phase	**[%] w/w**
Sorbitol (70% aquoeus solution)	21
Guar gum	0.15
Sodium saccharin	0.27
Sodium monofluorophosphate	1.1
CMC	1
Sodium bicarbonate	0.1
Sodium carbonate	0.4
Silica	2
Calcium carbonate	42
Titanium dioxide	0.1
Sodium lauryl sulfate	2
Methylparaben	0.1
Propylparaben	0.02
Water	28.7533
Flavor	1
CI Pigment Green 7	0.0067

guar gum provided three-dimensional structure to the toothpaste compositions.

In summary, it has been found that the binder system of CMC and guar gum provided in the calcium carbonate-containing toothpaste compositions produced according to the procedure described above, a smooth composition could be obtained, which had the desired viscosity and rheology and also solved the problem of stringiness and tailing in known toothpaste formulations (22).

The composition shown in Table 4.9 was employed to make a striped toothpaste, forming a first white phase. The second colored phase was comprised of the composition shown in the section *Ingredient Colored Phase* in Table 4.9.

The stripe quality was tested by extruding the striped toothpaste and quantitatively evaluating the stripe quality according to a stripe quality index numerical scale where 1 represents the worst stripe quality and 5 represents the best stripe quality. Ten data points were employed for the testing of each toothpaste, each data point corresponding to a respective property of a respective extrusion (A=stripe definition, B=stripe consistency of the first short ribbons during extrusion, C=stripe definition, D=stripe consistency of the first long ribbons during extrusion, E=stripe definition, F=stripe consistency of the second short ribbons during extrusion; G=stripe definition, H=stripe consistency of the second long ribbons during extrusion, I=stripe definition, J=stripe consistency of the last short ribbons during extrusion) (22). The results are shown in Table 4.10.

4.3.2 *Non-fluorinated Compositions*

Fluoride is typically added to toothpaste in order to reduce the decay of the teeth (23). In particular, fluoride incorporates itself into tooth enamel to make teeth more resistant to acids produced by plaque bacteria, as well as acids found in fruit juices, soda and certain foods.

Toothpastes that contain fluoride harden tooth enamel to make the entire tooth structure more resistant to decay and promote remineralization, which aids in repairing early decay. In a toothpaste, fluoride is commonly found in the form of sodium monofluorophosphate, stannous fluoride, or sodium fluoride.

Due to the toxicity of fluoride, the Food and Drug Administration (FDA) regards any toothpaste containing fluoride as a drug.

Table 4.10 Stripe quality index Example 1 (22).

Property	Value
Stripe definition	3.0
Stripe consistency of the first short ribbons during extrusion	2.5
Stripe definition	5.0
Stripe consistency of the first long ribbons during extrusion	5.0
Stripe definition	5.0
Stripe consistency of the second short ribbons during extrusion	5.0
Stripe definition	5.0
Stripe consistency of the second long ribbons during extrusion	5.0
Stripe definition	0.4
Stripe consistency of the last short ribbons during extrusion	2.0
Average	4.1

Accordingly, the FDA requires a warning on the label of any toothpaste containing fluoride stating: "If you accidentally swallow more than used for brushing, seek professional help or contact a poison control center immediately."

Moreover, the American Dental Association (ADA) requires that toothpaste manufacturers include the following language on all ADA approved toothpastes containing fluoride: "Do not swallow. Use only a pea-sized amount for children under six. To prevent swallowing, children under six years of age should be supervised in the use of toothpaste."

A non-fluoridated toothpaste has been described that is a significant source of at least one dietary supplement, such as a vitamin or a mineral. In particular, the non-fluoridated toothpaste is formulated so that it does not include fluoride and therefore is not considered to be a drug under the classifications set by the FDA. So, the non-fluoridated toothpaste is safe to swallow. The non-fluoridated toothpaste formulation contains at least one dietary supplement that when ingested supplements the diet of a mammal (23). Such a toothpaste composition can be made with ingredients in a range as shown in Table 4.11.

Table 4.11 Toothpaste composition (23).

Component	Contents range/[% w/w]		
Gum base	0.1	—	1.0
Sodium saccharin	0.20	—	0.40
Ethylene diamine tetraacetic acid	0.05	—	1.5
Sodium benzoate	0.20	—	0.40
Stevia	0.3	—	0.4
Xylitol	5.0	—	7.0
Polymer of ethylene oxide	0.02	—	0.06
Glycerin	26.0	—	28.0
Propylparaben	0.05	—	0.15
Titanium dioxide	0.4	—	0.6
Sorbitol	12.0	—	14.0
Niacinamide	0.07	—	1.0
D-Calcium pantothenate	0.03	—	0.06
Manganese chloride	0.025	—	0.04
Zinc lactate	0.2	—	0.40
Magnesium sulfate	0.6	—	0.9
Sea salt	0.05	—	0.15
Tetrasodium pyrophosphate	0.6	—	0.8
Amorphous silica	4.0	—	6.0
Dicalcium phosphate dihydrate	1.5	—	2.0
Amorphous silica 8.0 μ–11μ	8.0	—	10.0
Spearmint oil	0.5	—	1.5
Sodium lauryl sulfate	1.0	—	1.5
Sodium selenate	6.0×10^{-4}	—	9.0×10^{-4}
Vitamin A	2.5×10^{-5}	—	2.5×10^{-3}
Vitamin B1	2.5×10^{-5}	—	2.5×10^{-3}
Vitamin B6	2.5×10^{-5}	—	2.5×10^{-3}
Vitamin B12	2.5×10^{-5}	—	4.0×10^{-5}
Vitamin D3	2.5×10^{-4}	—	5.0×10^{-4}
Vitamin E	5.0×10^{-4}	—	2.0×10^{-3}

A non-fluoridated toothpaste has been developed that is enriched with a dietary supplement (11). The term *dietary supplement* was defined in the Dietary Supplement Health and Education Act (D-SHEA) of 1994. A dietary supplement is a product taken orally that contains a dietary ingredient intended to supplement the diet. The dietary ingredients may include vitamins, minerals, herbs or other botanical-based compounds, amino acids, and substances such as enzymes, organ tissues, glandulars, and metabolites.

Dietary supplements can also be extracts or concentrates, and may be found in many forms such as tablets, capsules, softgels, gelcaps, liquids, or powders.

The DSHEA places dietary supplements in a special category under the general umbrella of foods, not drugs. In particular, if a product contains less than 2% of the reference daily intake of a given dietary supplement, that product is not a *significant source* of that dietary supplement.

A non-fluoridated toothpaste enriched with at least one dietary supplement comprises a gum base, a thickening agent, at least one dietary supplement selected from the group consisting of vitamin B1, vitamin B6, vitamin A, vitamin D3, vitamin E, niacinamide, vitamin B12, D-calcium pantothenate and mixtures thereof.

In addition, the non-fluoride toothpaste also comprises at least one component selected from the group consisting of sodium selenate, manganese chloride, zinc lactate, magnesium sulfate, sea salt, tetrasodium pyrophosphate and mixtures thereof. The non-fluoridated toothpaste may further comprise a preservative and/or a flavoring agent. The components that can be used for such a composition are shown in Table 4.12.

Gum base is included to thicken the final toothpaste product so as to result in a paste, rather than a liquid. Gum base may be prepared by dissolving carboxymethyl cellulose in hot water, adding xanthan gum and mixing well.

The preparation of a formulation of a non-fluoridated toothpaste runs as follows (11):

Preparation 4–1: First component (B) is dissolved in component (A) in a container to form AB. Propylparaben is dissolved in the glycerin from component (C) in another container. Next, titanium dioxide is added and mixed. The mixture including propylparaben, glycerin, titanium dioxide,

Table 4.12 Ingredients for a non-fluoridated composition (11).

Component	Examples
A	Gum base: chicle, xanthan gum, guar gum, or a poly(butene)
B	Sodium saccharin, ethylenediamine tetraacetic acid (EDTA), sodium benzoate, stevia, xylitol, a polymer of ethylene oxide
C	Glycerin, propylparaben, titanium dioxide, sorbitol
D	Vitamin B1, vitamin B6, vitamin A, vitamin D3, vitamin E, niacinamide, vitamin B12, D-calcium pantothenate
E	Sodium selenate, manganese chloride, zinc lactate, magnesium sulfate, sea salt, tetrasodium pyrophosphate
F	Precipitated amorphous silica, dicalcium phosphate dihydrate, precipitated amorphous silica
G	Natural spearmint oil, sodium lauryl sulfate

and any other components of component (C), save sorbitol, is then transferred to the container containing AB. A portion of the sorbitol solution from component (C) is used to rinse titanium dioxide and/or glycerin from the container, and is transferred to the container containing AB. The remaining sorbitol solution is then added to form ABC. Next, component (D) is added to ABC, dissolved, and mixed to form ABCD. Component (E) is added to ABCD and mixed to form ABCDE. Next, component (F) is added to ABCDE to form ABCDEF. The spearmint oil from component (G) is then added to ABCDEF and mixed. Next, the beads containing at least one dietary supplement are added to the mixture of ABCDEF and spearmint oil, and mixed gently. Lastly, sodium lauryl sulfate is added and mixed to form ABCDEFG. ABCDEFG should be mixed gently following the addition of the beads containing at least one dietary supplement and/or sodium lauryl sulfate to avoid breaking the beads and/or foaming.

4.3.3 *Alkyl Sulfate and Orthophosphate Free Composition*

A special dentifrice composition for combating the problem of dental erosion and tooth wear has been developed (24). The composition contains a fluoride ion source and a silica dental abrasive. The composition has a relative dentine abrasivity (RDA) value from 20 to 60 and a pH in the range 6.5 to 7.5.

It is free from an orthophosphate buffer or a water-soluble salt of a C_{10} to C_{18} alkyl sulfate. The usage of an orthophosphate buffer should be avoided, as its inclusion reduces the fluoride uptake into the enamel. Further, sodium lauryl sulfate has good detergent and cleaning properties, but it can cause the formation of an insoluble potassium lauryl sulfate precipitate (24).

These compositions are particularly beneficial in combating dental erosion, since not only are they able to harden and thereby protect teeth from an acidic erosive challenge, but also are able to reharden enamel softened by an acidic erosive challenge (24).

Because of their low abrasivity, substantially neutral pH and the absence of a powerful C_{10} to C_{18} alkyl sulfate surfactant, such as sodium lauryl sulfate, commonly used in oral compositions, these compositions do not exacerbate the problems associated with dental erosion and tooth wear.

In order to treat dental hypersensitivity, the composition further contains a desensitizing agent. Examples of desensitizing agents include a tubule blocking agent or a nerve desensitizing agent described (25).

Preferred desensitizing agents include a strontium salt such as strontium chloride, strontium acetate or strontium nitrate or a potassium salt such as potassium citrate, potassium chloride, potassium bicarbonate, potassium gluconate and in particular potassium nitrate (24).

4.3.3.1 *Acidulating Agents for Preparing a Silica Product*

Precipitated silica can be prepared by adding an acidulating agent to an alkali metal silicate to precipitate amorphous silica (26). The resulting precipitate is usually filtered away from the reaction medium and subsequently washed and dried. Typically, the dried silica is then mechanically comminuted in order to provide a suitable particle size and size distribution.

On an industrial scale, silica can be prepared by a step-wise batch process that incorporates the aforementioned steps. The equipment needed for such a process can be capital intensive and often leads to inefficiency in the process, particularly when idle time exists when reactants are not being consumed. While various other silica production processes exist, many of these processes are difficult to control and scale-up, and many still require extensive processing steps after the silica has been prepared (26).

A continuous process for preparing a silica product has been reported. This process runs as (26):

1. Continuously feeding an acidulating agent and an alkali metal silicate into a loop reaction zone comprising a stream of liquid medium; wherein at least a portion of the acidulating agent and the alkali metal silicate react to form a silica product in the liquid medium of the loop reaction zone,
2. Continuously recirculating the liquid medium through the loop reaction zone, and
3. Continuously discharging from the loop reaction zone a portion of the liquid medium comprising the silica product.

The silica products are particularly useful in dentifrice compositions as part or all of the abrasive or cleaning agent. As used herein, a dentifrice composition refers to a composition that can be used to maintain oral hygiene, for example, by cleaning accessible surfaces of the teeth.

The dentifrice composition can be a liquid, powder, or paste. Typically, the dentifrice compositions are primarily composed of water, detergent, humectant, binder, flavoring agents, and a finely powdered abrasive. The silica particles, when incorporated into dentifrice compositions, can be present at a level of from about 10% to about 35% by weight (26).

4.3.3.2 *Tartar Control Agents*

Tartar control (anticalculus) agents can be additives for toothpaste compositions (27). Examples of tartar control agents are shown in Table 4.13.

Testing of Tartar Control Agents. The effects of anticalculus dentifrices were compared with other commercially available dentifrices using vitro models of dentin sensitivity (28). The changes in the hydraulic conductance of dentin discs were measured with and

Table 4.13 Tartar control agents (27).

Compound
Pyrophosphates
Polyaminopropanesulfonic acid
Polyolefin sulfonates
Polyolefin phosphates
Azacycloheptane-2,2-diphosphonic acid
N-Methyl azacyclopentane-2,3-diphosphonic acid
Ethane-1-hydroxy-1,1-diphosphonic acid
Ethane-1-amino-1,1-diphosphonate
Phosphonoalkane carboxylic acids
Monobasic, dibasic and tribasic sodium phosphates
Sodium tripolyphosphate
Tetrapolyphosphate
Mono-, di-, tri- and tetrasodium pyrophosphates
Sodium trimetaphosphate
Sodium hexametaphosphate

without a smear layer before and after treatment and also after a post-treatment acid etch.

The capacity of dentifrices to occlude open dentinal tubules *in-vitro* was also assessed by a scanning electron microscope (SEM). A good correlation between the test performed in this study and values reported in the literature were found. Tartar control dentifrices gave reductions in fluid flow rates through the dentin discs comparable to those obtained with Promise, Sensodyne, Thermodent and Denquel toothpastes. Additionally, tartar control dentifrices did not remove microcrystalline debris, i.e., smear layers, from the surfaces of dentin *in-vitro*. These results were also confirmed by SEM. Thus, according to the hydrodynamic theory of dentin sensitivity, these *in-vitro* results suggest that pyrophosphate-containing dentifrices should reduce dentinal sensitivity (28).

In a clinical trial a test was performed by 92 persons (29). The purpose of the trial was to evaluate the effects of toothpastes, in varying concentrations of flavor and tartar control agents. Four formulations of toothpastes were assessed (29):

1. A control-low flavor with no tartar control,
2. A medium flavoring with medium tartar control,
3. A high flavoring with medium tartar control, and

4. A medium flavoring with no tartar control.

The soft-tissue reactions were assessed objectively and independently by three examiners. Subjective perceptions about each toothpaste were gathered by a structured, open-ended questionnaire. The tartar control toothpastes, 2 and 3, resulted in statistically significant ($p < 0.005$) higher rates of mucosal reactions, e.g., ulceration, sloughing, erythema, migratory glossitis, than the non-tartar control toothpastes 1 and 4 (29).

When the oral mucosal reaction rates were adjusted for multiple clinical observations within each subject at the same point in time there was no statistically significant ($p > 0.05$) difference between male (0.25) and female (0.28) subjects. In this study population, the order of preference was observed as toothpaste 1>4>2>3. The major reasons for disfavor were burning sensation in toothpastes 3 (2.4%) and B (4%) (29).

4.3.4 *Thin Film Toothpaste Strip*

Toothpaste is typically in the form of a tube containing a gel containing a multiplicity of ingredients (30). These tubes are heavy, and have a tendency to leak or dry out if the cap is not secured tightly. Additionally, toothpastes contain fluoride compounds that present risks of overdelivery of active agent due to the inability to accurately measure and deliver toothpaste portion.

Tooth decay is an infectious, multifactorial disease with global reach. Good dental practices have made a huge impact in reducing risk factors associated with poor dental hygiene. The introduction and acceptance of fluoride has led to a reduction of the incidence of dental caries and has been demonstrated to slow or reverse the progression of existing lesions, i.e., prevents cavities. The ability of fluoride to inhibit or even reverse the initiation and progression of dental caries is well documented.

The first use of adjusted fluoride in water for caries control began in 1945 and 1946 in the United States and Canada, when the fluoride concentration was adjusted in the drinking water supplying four communities. The success of water fluoridation in preventing and controlling dental caries led to the development of fluoride-containing products, including toothpaste, i.e., dentifrice, mouth rinse, dietary supplements, and professionally applied or prescribed gel, foam, or varnish. Unfortunately, the success of fluoride use has led to over use in the population (30). Fluoride intake particularly among children aged 6 years and younger has been on the rise, increasing the risk for enamel fluorosis.

Thus there exists a need for alternative dosage forms to tubes of gel like toothpaste tubes and containers (30). A non-adhering oral tape, film or strip home oral care composition comprising a cleansing effective amount of a teeth cleaning agent and a fluoridating effective amount of a fluoridating agent has been developed.

Also, methods of forming and using these non-adhering oral tapes, films or strips, and particularly to non-adhering oral tapes, films or strips containing precise dosage amounts of fluoride, have been shown (30). A precision fluoride dosage can help to mitigate the problems of fluoride overdosage associated with standard dentifrice formulations.

The production of a thin film toothpaste strip runs as follows (30): Hydroxymethyl cellulose extruded as a thin film layer is coated with an extruded formulation containing sodium fluoride, hydrated silica, triclosan, sodium lauryl sulfate, and a copolymer of methyl vinyl ether and maleic anhydride and dried in an oven to form a non-adhering oral tape, film or strip home oral care composition.

4.3.5 *Enamel Protectant and Repair Toothpaste*

Substantially aqueous-free, enamel protectant and enamel repair toothpastes have been described that are containing: Stannous fluoride, calcium and a substantivity agent which is an emulsion of poly(dimethylsiloxane) in a nonionic surfactant (31).

The toothpastes contain stannous fluoride from between 850 *ppm* and about 1500 *ppm*, in an aqueous-free, substantivity agent (32). The toothpaste treatments can protect and repair enamel more effectively than toothpastes and toothpaste treatment containing comparable or substantially higher levels of fluoride, as indicated by comparative enamel protectant factor and enamel repair factor values.

The stannous fluoride and calcium in biofilm present on enamel is enhanced through calcium-binding shifting from bidentate to monodentate in the presence of stannous fluoride. Toothpastes show substantially improved, enamel protectant factor and enamel repair factor values compared to fluoride brushing treatments with comparable or higher fluoride levels (31).

4.3.6 *Striped Toothpastes*

4.3.6.1 *Striped Toothpaste Stable to Color Bleeding*

A major problem impacting the aesthetic appearance of a striped toothpaste is the bleeding or migration of color from one component into another (33). This problem is especially severe if one colored component is applied to the surface of a white base. For this reason, a colorant that exhibits substantially no visible bleeding is required.

Striped dentifrice products containing water-soluble dyes have been described (34–36). A disadvantage to the use of water-soluble dyes enumerated in these patents is that a visible bleeding is observed (33).

Dentifrices containing encapsulated ingredients have been disclosed (37–40), such as flavors, whereby such ingredients are maintained substantially separate from other dentifrice ingredients during manufacture and storage, while subsequently releasing the encapsulated ingredients into the dentifrice during tooth brushing.

It is also known (19) that water-insoluble dyes can be encapsulated in capsules wherein the shell material is formed from nontoxic naturally occurring waxes such as carnauba wax, candelella wax, castor wax, paraffin wax and bayberry wax. Although the encapsulation of the dyes in these waxes overcome dye migration to some extent, undesirable levels of bleeding persist during accelerated aging conditions, as, for example storage, of the striped dentifrice for 4–6 weeks at 50°C (33).

An aesthetically pleasing, pleasant tasting, substantially nonbleeding, striped dentifrice composition has been developed that contains two dentifrice components wherein at least one of the components is a paste or gel containing a colorant entrained in a matrix of a high density poly(ethylene) (HDPE) having a melting point range, as determined by differential scanning calorimetry (DSC), between about 110°C and about 140°C.

In particular, by analyzing the data on the temperature increase observed by subjecting a thermoplastic material to DSC, it is possible to observe a state change of the material under heat application and heat absorption peaks accompanying a phase transition and melting of the thermoplastic material.

The HDPE entrained colorants are unexpectedly substantially non-bleeding when present in conventional toothpaste or gel formations, in particular when contrasted with similar colorants entrained in wax and synthetic polymeric resins including paraffin wax and low density poly(ethylene) (LDPE) (33).

It is believed that the molecular weight, degree of crystallinity, and degree of branching groups of HDPE as reflected by its melting point are responsible for the superior non-bleeding properties of colorants entrained in this polymer. For example, HDPE is more crystalline in nature and has a lesser degree of branching than LDPE and paraffin waxes and is significantly superior to these latter materials in the preparation of non-bleeding colorants for striped toothpaste products (33).

Colorants suitable for entrainment or encapsulation in the HDPE matrix are physiologically compatible water-soluble dyes and lakes including natural or synthetic dyes of the types permitted in foods and drugs, such as those listed in Title 21 of the U.S. Code of Federal Regulations, section 74 (41). In addition to these water-soluble dyes, it is also possible to use water-insoluble dyes. Examples are shown in Table 4.14.

Table 4.14 Physiologically compatible dyes (33).

Water-soluble	Water-insoluble
FD&C Blue #1	Eyeshadow Blue KO
FD&C Yellow #10	Colour Index 77 510
	EG-No. Blue 15 (C-Blue 17)

Also, mixtures of water-insoluble dyes and water-soluble dyes can be used, for example, Eyeshadow Blue KO and Lemon Yellow ZN 3, in which case green hues are obtained. Preferred colorants are used in amounts of preferably 7–30%.

Preferred lakes are those certified by the Color Certification Laboratory of the Food and Drug Administration of the Health, Education and Welfare Department of the United States Government, for example, FD&C Blue No. 1 Lake, FD&C Blue No. 2 Lake, FD&C Red No. 3 Lake, FD&C Yellow No. 5 Lake, and FD&C Yellow No. 6 Lake (33,42).

4.3.6.2 Striped Toothpaste with Calcium Carbonate

Various binder components, and binder systems incorporating combinations of binder components, are known in the art of making toothpaste (43). Different binder components impart different rheological properties to the toothpaste. Cellulose polymers, in particular CMC, are widely used in toothpastes to act as a binder and thickener.

However, for some toothpaste formulations, in particular toothpaste formulations containing a significant proportion of calcium carbonate particles, the addition of such a cellulose polymer can cause the toothpaste to exhibit a *stringy* effect and to exhibit *tailing* when the toothpaste is manufactured. Such a stringy effect and tailing are manifested, for example, when the paste is mixed using a mixing blade and, after the blade has been lifted from the mixer tank, paste on the blade falls downwards under gravity as tails rather than being bound to the blade as a single non-flowing paste body.

Filling the toothpaste package effectively and reliably is a problem if the toothpaste exhibits the stringy effect and tailing, particularly for toothpaste formulations containing a significant proportion of calcium carbonate particles.

Therefore, there is a need for a toothpaste composition which contains calcium carbonate particles and exhibits good rheological properties, in particular reduced stringiness and reduced tailing (43).

In striped toothpaste formulations containing a significant proportion of calcium carbonate particles, it is known to use a binder system which controls the stripe quality. In particular, is it is known to use a binder system incorporating the combination of CMC and MAS to provide a high stripe quality.

However, the use of MAS suffers from the problem that there have been only a few qualified suppliers of toothpaste-grade MAS, which increases the cost and complexity of the supply chain.

A toothpaste composition has been described which contains calcium carbonate particles and exhibits good rheological properties, in particular a reduced stringiness and a reduced tailing. Such toothpaste compositions are shown in Table 4.15.

Table 4.15 Toothpaste compositions (43).

Ingredient	Amount/[%]
Sorbitol (70 wt% aquoeus) solution	21.0
Guar gum	0.10
Xanthan gum	–
MAS	–
Sodium saccharin	0.27
Sodium monofluorophosphate	1.10
CMC	1.0
Sodium bicarbonate	0.1
Sodium carbonate	0.4
Silica	2.0
Calcium carbonate	42.0
Titanium dioxide	0.1
Sodium lauryl sulfate	2.0
Methylparaben	0.1
Propylparaben	0.02
Deionized water	28.81
Flavor	1.0
Total	100

The replacement of MAS with guar gum provided a good stripe quality. In contrast, replacing MAS with xanthan gum, provided a poor stripe quality, with a poor separation between the white phase and the colored stripe (43).

4.3.7 *Color Changing Compositions*

It is recommended that children should brush their teeth for at least 45–60 s, and adults for at least 90–120 s (44). Most people, especially children, do not brush their teeth for a sufficient period of time to obtain maximum benefit; moreover, they have difficulty accurately estimating the time necessary to brush.

Oral care compositions have been developed that contain a dissolvable film with a pigment, which provides a color change signal after a sufficient period of brushing (44). It has been discovered that the timing of the color change can be adjusted through the film by both composition and thickness to occur after a certain time so as to provide a visual signal to the consumer that enough time has been spent brushing.

A special challenge is how to stabilize the films so that they do not dissolve in the dentifrice but dissolve only when exposed to water and brushing action (44). In a typical toothpaste, such films break down within 2–4 *d*, leaving the pigment to bleed within the dentifrice. Toothpaste formulas with a water activity above 0.78 show a significant film instability upon aging. However, compositions having water activity below 0.78 are stable over two months.

Using an experimental design approach, various factors have been evaluated for their impact on film stability. Higher levels of anionic polymer in the formulation enhance the stability. On the other hand, anionic surfactants, such as sodium lauryl sulfate, can have an unexpected detrimental effect on stability, and the levels must be controlled to ensure an acceptable stability (44).

A prototype film was developed by encapsulating a pigment into a dissolvable polymer film. During brushing, the films swell from water and disintegrate, releasing the pigment and, thus, color change occurs that lets the consumer know when the brushing is done.

One use for this film is for incorporation into a clear gel toothpaste providing a color change signal to the consumer after a predetermined brushing time. The ingredients for the prototype film are shown in Table 4.16.

Table 4.16 Composition with a pigment in a dissolvable polymer film (44).

Ingredient	[%] per weight
Hydroxypropylmethyl cellulose	48.2
Pigment	30
Propylene glycol	18
Polysorbate (Tween 80)	3.8
Total Amount	100.0

Another earlier published method to achieve a color change is to use color reagents with two separate color components (45). The components are adapted to interact after a predetermined brushing time to produce a third color.

It has been found that a non-foaming, color-changing toothpaste can be formulated by incorporating color reagents for interacting to provide a color change indicative of tooth brushing time. The reagents contain two separate color components selected from the group consisting of FD&C dyes, FD&C lakes (46) and natural food colors (45).

4.3.8 *Two Phase Toothpaste*

Oral care compositions are used for a wide variety of purposes, including for enhancing personal health, hygiene, and appearance, as well as for preventing or treating a variety of diseases and other conditions in humans and in animals (47).

Two phase dentifrice compositions have been described. Such oral care compositions contain (47):

1. A first phase comprising clinically efficacious 2,4,4′-trichloro-2′-hydroxydiphenyl ether, cf., Figure 4.6, admixed in a first orally acceptable aqueous vehicle; and
2. A second phase with a stannous salt admixed in a second orally acceptable aqueous vehicle, wherein
3. The second phase contains not less than 10 mol% of soluble stannous ion respective to a mathematical sum of moles of the soluble stannous ion and moles of the stannous salt in the second phase.

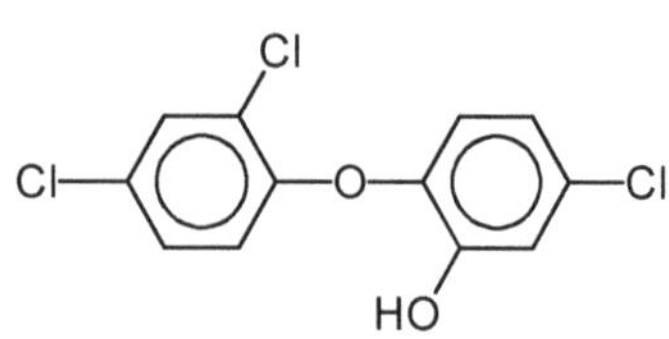

Figure 4.6 2,4,4′-Trichloro-2′-hydroxydiphenyl ether.

The basic components used in a two phase composition are shown in Table 4.17. Several examples of more detailed compositions have also been reported (47).

The above-described compositions minimize the astringent taste traditionally associated with stannous oral care compositions and incorporate high cleaning silica to reduce tooth staining.

Also, it has been found that the addition of castor oil and sodium bicarbonate contribute significantly to the improvement of the taste. In the dual stream delivery system afforded by the dual-compartment tube, one side of the system delivers the stannous ingredient and the second side delivers tartar control benefits along with flavorants that successfully mask the stannous astringency (47).

4.3.9 Composition with Diamond Particles

Toothpastes are used to clean, bleach, whiten, and otherwise treat the teeth. The active ingredients in such a composition are contained within a carrier (48).

Most toothpastes contain various types of silica to debride and to scrub the external surface of the teeth. Scrubbing removes the organic film formed of salivary proteins which covers the teeth and which is known to become stained and discolored by foods, such as coffee, tea and berries, as well as by tobacco smoke, cationic antibacterials, and chromogenic bacteria (48).

The physical removal of the stained pellicle is a simple and effective means of removing the undesirable surface staining and discoloration which occurs daily. Furthermore, such a physical removal of the pellicle also removes plaque bacteria on the pellicle surface.

The carrier can be a toothpaste or gels, i.e., brushing gels and bleaching gels. Gels are thickened by a gelling agent through hydrogen bonds in a dispersion medium to produce a semisolid, transparent, jelly-like material. In contrast, pastes are thickened by the addition of fillers.

Also, combinations of such ingredients have been described. These cleaning compositions are useful and satisfactory for manual brushing of teeth. However, these conventional agents are not satisfactory for use by cosmetic dental patients with porcelain veneers, caps or crowns. Harder and finer abrasives are necessary to effect a higher degree of polishing as opposed to mere cleaning of such surfaces to provide a smooth, mirror-like finish (48).

Table 4.17 Two phase composition (47).

Components (first phase)	Contents range/[% w/w]		
Synthetic glycerin	8	—	25
Sodium carboxymethyl cellulose	0.5	—	1.5
Iota carrageenan gum	0.2	—	1.2
Sodium saccharin	0.0001	—	0.8
Titanium dioxide		—	1.5
Sorbitol	8	—	25
Methyl vinyl ether/maleic anhydride copolymer	0.5	—	8
Sodium hydroxide for pH of 7–9		—	
Silica	8	—	25
Sodium bicarbonate		—	5
Flavor oil	0.5	—	2
Propylene glycol	0	—	3
2,4,4′-Trichloro-2′-hydroxydiphenyl ether	0.2	—	1
Sodium lauryl sulfate	0.5	—	4
Sodium fluoride	0.243	—	0.486
Components (second phase)	**Contents range/[% w/w]**		
Anhydrous citric acid	0.05	—	1
Trisodium citrate dihydrate	0.5	—	5
Stannous fluoride	0.05	—	2
Stannous chloride dihydrate	0.25	—	5
Synthetic glycerin	10	—	40
Sodium carboxymethyl cellulose gum	0.5	—	1.5
Xanthan gum	0.2	—	1
Sodium saccharin	0.001	—	0.8
Tetrasodium pyrophosphate	0.005	—	1
Poly(oxypropylene)-poly(oxyethylene) block copolymer	0.2	—	4
Hydrogenated castor oil	0.05	—	10
Color dye	0.0005	—	2
Sodium lauryl sulfate	0.5	—	4
Flavor oil	0.5	—	2
Silica	8	—	35

A toothpaste composition has been described that is useful for cleaning cosmetic dental surfaces such as porcelain, composites and also natural teeth (48). The composition contains a diamond abrasive, or diamond particulates, such as diamond dust.

It has been discovered that a dentifrice or polishing composition comprising diamond particulates can be prepared that will maintain the luster and gloss of porcelain dental work with a low abrasion on natural tooth enamel. The composition may include many other compounds that are useful in toothpastes and gels, such as flavoring agents, thickeners, stabilizing agents, colors, and humectants.

The diamond particles or diamond dust included in the toothpaste are present in a size range of about 0.01 μm to about 5.0 μm in diameter. The diamond particulates can be from any commercially available source. They are typically sold in various sizes or meshes and are used primarily in the grinding and polishing arts. The diamond powder should have a superabrasive quality.

In addition, commercially available abrasives may be used in combination with the diamond powder. These abrasives include precipitated silicas having a mean particle size of up to about 20 μm, insoluble phosphates, such as dicalcium orthophosphate dihydrate, calcium pyrophosphate, β-calcium pyrophosphate, tricalcium phosphate, calcium poly(metaphosphate) and insoluble sodium poly-(metaphosphate).

A dentifrice composition with diamond particles is shown in Table 4.18.

Table 4.18 Composition with diamond particles (48).

Ingredient	[%] per weight
Water	10 - 70
Xanthan gum	0.1 - 10
Sorbitol	5.0 - 30
Glycerin	5.0 - 30
Plant extract	0.1 - 5.0
Xylitol	0.05 - 5.0
Sodium monofluorophosphate	0.1 - 2.0
Silica-SIDENT 9 (Degussa)	0.1 - 20.0
Silica-SIDENT 22 S (Degussa)	0.1 - 20.0
Titanium dioxide	0.1 - 5.0
Diamond Powder MSY 0.5 μm	0.001 - 1.0

A composition that is useful for patients with sensitive teeth is shown in Table 4.19.

Table 4.19 Composition for patients with sensitive teeth (48).

Ingredient	[%] per weight
Water	23.49
Xanthan gum	1.00
Sorbitol	20.00
Glycerin	21.00
Plant extract	0.50
Xylitol	0.20
Sodium monofluorophosphate	0.70
Silica-SIDENT 9 (Degussa)	6.00
Silica-SIDENT 22 S (Degussa)	13.00
Titanium dioxide	1.00
Diamond Powder MSY 0.5 μm	0.01
Cocamidopropyl betaine	2.00
Aloe vera oil gel	0.10
Mint fragrance	1.00
Potassium nitrate	5.00

4.3.10 *Compositions with Improved Rheology*

It is believed that the water activity is an important property of oral care compositions (21). The water activity can affect the solubility of the ingredients, which in turn can be used to modify the product rheology.

The water activity refers to the equilibrium relative humidity or partial vapor pressure of the air above a product sample in a closed system divided by the relative humidity of air above pure water at the same temperature. The water activity can be expressed in percents of the equilibrium relative humidity.

Compositions that contain the same total concentration of water can have water activity values that differ dramatically depending on the hygroscopic characteristics of the other ingredients in the compositions. Nearly all paste or gel semisolid formulations use polyhydric alcohols to help in limiting the water loss over the self life of the product.

Many polyhydric alcohols are classified as humectants and the most commonly used humectant materials in dentifrice formulations are glycerin, sorbitol, propylene glycol, poly(ethylene glycol)s and mixtures from these compounds.

Oral care compositions with an improved rheology have been developed. The compositions contain humectants such as PEG (21).

4.3.11 *Toothpaste Composition with Improved Shelf Life*

It has been found that PEG improves the shelf life and stability of a toothpaste composition that contains semi-refined carrageenans (49). The properties of carrageenans are documented in Section 4.3.11.1, just below.

The following procedure was used to prepare toothpaste batches under the hot process in order to demonstrate the effectiveness of the poly(ethylene glycol) in improving stability in a toothpaste that is contains a semi-refined carrageenan.

The specific toothpaste formulations were a chalk-based toothpaste wherein each binder comprised 36% conventional extract iota carrageenan and 50% semi-refined iota carrageenan. The balance of the binder was silica in an amount of 14%. The conventional extract iota carrageenan that was used in the testing below was the same and had a viscosity of about 24 *cps*. The semi-refined carrageenan samples used were made using similar processes with minor variations.

In general, the process that was used was (49):

1. Raw seaweed was washed with an aqueous 1.5% NaCl solution,
2. The seaweed was modified by treating such with an aqueous NaOH and NaCl solution at a temperature of about 45°C,
3. The seaweed was washed twice at a pH of about 8,
4. The seaweed was then dried and ground, and
5. Either alcohol washed (three times with increasing amounts of isopropanol) or ion exchanged, in EDTA/NaCl/alcohol.

4.3.11.1 *Carrageenan*

Carrageenan is a complex mixture of sulfated polysaccharides with linear polymers of 1,3-α-*D*-galactose units and 1,4 linked β-*D*-galactose units (49).

Carrageenans have curl forming helical structures (50). This gives them the ability to form a variety of different gels at room temperature. They are widely used in the food and other industries as thickening and stabilizing agents. Various carrageenan structures have been shown (50).

Carrageenan is produced by red seaweeds, where it functions as the principle structural polysaccharide (49). It is located within the cell wall and intracellular matrix of the plant tissue. The carrageenan content of commercially harvested seaweeds is generally 30–80%, based on the seaweed dry weight.

Carrageenan finds a wide applicability as a food ingredient and is functional in foods such as dairy products, water dessert gels, meat products, confections, beverages, dressings and other such products. Carrageenan is also useful in products such as cosmetics, toothpaste and other personal care products, in soft gel capsules, and in other industrial, medical, pharmaceutical, and agricultural applications. The molecular weight of carrageenan products is typically from about 100 $k\,D$ to about 1,000 $k\,D$.

Carrageenans have the ability to form an almost infinite variety of gels at room temperature with a variety of gelling and melting points. Carrageenan solutions can thicken, suspend, and stabilize particulates, colloidal dispersions and water/oil emulsions. The solutions shear thin, which allows them to be pumped easily. Also, the sheared solutions rapidly rebuild viscosity and suspending power upon standing. Depending upon the application, carrageenans present in parts per million up to a few percent by weight provide gelling, thickening, suspending, binding and/or generates a desired product feel or texture.

Carrageenan is generally soluble in warm water, in which it forms a viscous solution. It is insoluble in most organic solvents and typically forms complexes with proteins.

In a typical process for producing refined carrageenan, crude seaweed is first washed with cold water or seawater to remove sand and other particles that may be present after seaweed has been harvested (49). Carrageenan typically does not swell during the cold wash, primarily because carrageenan in seaweed is associated with structural components of the seaweed, generally cellulose.

Depending upon the seaweed species, following the cold wash, a hot water extraction process is typically performed in which the extracted carrageenan is treated with an aqueous base at high temperature. Generally, the base used is an alkali or alkaline earth metal hydroxide such as, for example, sodium hydroxide, calcium hydroxide, or potassium hydroxide.

This high temperature aqueous base modification step leads to the formation of 3,6-anhydro linkages in the galactose units of the carrageenan. After this modification, the hot extract is filtered to remove insoluble materials such as cellulose, hemicelluloses and other particulates and acid is added to adjust the pH to 7.5 to 10.5.

The filtrate can then be concentrated to about 4% carrageenan for further processing. Optional process steps after extraction include centrifugation and bleaching. Refined carrageenan is typically obtained by precipitation of the extract from the aqueous solution with potassium chloride or an alcohol such as isopropanol. The resulting carrageenan product is subsequently dried and ground (49).

4.3.11.2 *Toothpaste with Low Levels of Carrageenan*

Many carrageenan-containing toothpastes are generally recognized as having very good properties such as acceptable physical stability, low stringiness and good rheology. Toothpastes with acceptable physical stability do not readily harden on the shelf and do not exhibit phase separation such as water or flavor separation (51).

Low stringiness is important not only to the consumer but also in high volume manufacturing where high speed filling lines require that the toothpaste ribbon cuts off sharply from the tube. Toothpastes having good rheology will be easy to dispense from the tube yet stand up well on the brush. Moreover, these properties may be provided in a carrageenan-containing toothpaste that has appealing taste, has good cleansing effect, is easy to rinse, has excellent mouthfeel, and has a smooth, pleasant appearance.

However, despite the well-known benefits of using carrageenan in toothpaste, the wider use of carrageenan has been limited by its high cost compared to other binders, especially CMC. Cost is a particular issue in parts of the world where toothpaste, despite its importance for dental hygiene, remains unaffordable.

A part of the higher cost of using carrageenan comes from the relatively high cost associated with obtaining the carrageenan raw material from its natural source, seaweed. Another part of the cost comes from the inherent problem of manufacturing toothpaste with binders that build viscosity.

For example, heat transfer becomes less efficient as formulations become more viscous, requiring greater time for heating and cooling. As formulations become more viscous they do not mix as readily and therefore require greater agitation to achieve appropriate mixing. Also, with high viscosity formulations it is more difficult to obtain consistent and accurate metering at the filling equipment. As a result of the mixing and pumping required for high viscosity formulations, most toothpastes today are prepared by a batch process (51).

Continuous processes, which are theoretically more economical than batch processes, are generally not feasible for toothpastes with carrageenan. However, there have been attempts to design a suitable continuous process as already previously described (52).

A process for making toothpaste with low levels of certain carrageenan types has been described (51). The process may be applied to making toothpastes having suitable viscosity using carrageenan levels as low as about 0.05%. The compositions contain one or more of the following carrageenan types: Iota carrageenan, modified kappa carrageenan, or mixtures from these compounds.

The modification of the carrageenan occurs during its processing and extraction from seaweed as a result of alkali treatment at elevated temperatures. A process for alkali modification has been described by Stanley in 1963 (53). Various alkaline materials may be used for the modification, including alkaline earth hydroxides such as the hydroxides of calcium, barium, and strontium, sodium carbonate, trisodium phosphate and sodium metaborate (51).

Preferred alkaline materials include calcium hydroxide by itself or a mixture of calcium hydroxide with sodium hydroxide. The pH during the modification is typically above about pH 9, preferably between about pH 9.5 and 13. The temperature for the alkali treatment may range from about 80°C to about 150°C, with preferable temperatures between 90°C and about 100°C.

An increase of the viscosity of at least about 100% can be observed when the compositions are allowed to quiescently cool from a temperature at or above about 45°C. Toothpastes that are prepared using this method are easier to process and require much less carrageenan relative to toothpastes with a similar viscosity that are prepared by conventional methods (51).

4.3.12 *Bleach-Stable Toothpaste*

Bleach-stable compositions and processes and methods of making such compositions have been reported (54). These compositions contain an effective amount of a discontinuous phase consisting of one or more hydrophobic, inert, bleach-stable abrasive agents that are polymers, copolymers and crosspolymers of granular nature of about between 2 μm and 150 μm particle size, present in the operative range of about 10% and 75% with respect to the total weight of the composition (54).

These abrasive agents are dispersed or otherwise distributed in a continuous phase with one or more bleaching actives emulsified in an inclusion complex polymer of delayed water and saliva solubility with said emulsified bleaching actives being capable of forming a long-lasting bleaching film on tooth surfaces and being present in the operative range of about 0.5% and 55% with respect to the total weight of the composition.

Examples of some inclusion complex polymers and copolymers to slightly retard the solubility of the bleaching actives are shown in Table 4.20.

When topically applied to tooth surfaces, the inclusion complex polymers and copolymers retard the activity of the bleaching agents sufficiently to allow the abrasive agents to remove the plaque and pellicles from enamel surfaces by mechanical action, thus allowing for easier diffusion of the bleaching agents into the plaque and pellicle-free enamel and allowing the bleaching active to effectively whiten the tooth surfaces. Examples of how to prepare such compositions are shown below:

Preparation 4–2: Bleaching composition (A): In 50 *g* purified water, 1.5 *g* of emulsifier Carbopol 934/Poly(vinylpyrrolidone) in 75:25 ratio are added and dissolved with gradual stirring. To the mixture 20 *ml* of hydrogen peroxide (50%) was added and mixed for an additional 5–10 *min*. The acid composition was then adjusted between pH 5.5 and 6.5 with 10% NaOH. The composition thickens to a gel and set aside.

Table 4.20 Inclusion complex polymers (54).

Compounds
Emulsifying waxes
Glycerol and glycol esters
Ethoxylated fatty esters
Block polyols
Ethoxylated alcohols
Sorbitan esters
Ethoxylated sorbitan esters
Organosilicones
Silicone fluids
Volatile silicones
Poly(ethylene glycol)
Poly(propylene glycol)
Ammonium styrene/acrylates copolymers
Acrylates/ammonium methacrylate copolymers
Poly(methacrylate)
Poloxamer
Carbopol
Poly(ethylene glycol)/polypropylene glycol
Vinyl pyrrolodone/acrylic acid copolymer

Abrasive suspension (B): Then, in a separate vessel 210 *g* of methyl methacrylate crosspolymer GMX-0610 obtained from Perspere Corp. was added.

Continuous phase (C): In another separate vessel a continuous phase was prepared with the ingredients shown in Table 4.21.

After the bleaching composition (A) has been prepared with the desired consistency, 50 *g* of this composition is added to 50 *g* of the water-insoluble abrasive suspension (B) and the intimate mixture of the two immiscible phases are dispersed in each other and then, with the aid of the colloidal mill, agitated until extremely fine homogeneous dispersion is obtained. Then 100 *g* of the dispersion so obtained was added to 50 *g* of the continuous phase (C) and the two phases mixed in a colloidal mill and the resultant composition comprised the discontinuous phases (A) dispersed homogeneously throughout the continuous phase (B) and (C).

Table 4.21 Continuous phase composition (54).

Compound	[%] w/w
Sodium fluoride	1.05
Propylene glycol	24.10
Sodium lauryl sulfate	5.04
Water	43.40
Vinylpyrrolidone/Acrylic acid	1.02
Hydroxyethyl cellulose	2.01
Glycerin	18.85
Sodium saccharin	0.47
Flavor	2.76
Sodium benzoate	0.55
Benzoic acid	0.06
Sodium EDTA	0.14
Sodium hydroxide (10% solution)	0.55

Preparation 4–3: The preparation 4–2 was repeated except that in phase (A) to 50 *g* of deionized water 28 *g* of carbamide peroxide in addition to 20 *ml* of hydrogen peroxide were added to the composition with constant stirring in the same manner as in preparation 4–2.

The advantage of having carbamide peroxide in this method is that it provides an excellent quick dissolution and the onset of bleaching activity due to its higher pH level than that provided by the hydrogen peroxide. Carbamide peroxide, when dissolved, breaks down into hydrogen peroxide and urea, which is an ammonium compound that is highly alkaline.

Also, a peroxygen/anesthetic tablet composition for the treatment of post-extraction septic sockets was made containing the ingredients shown in Table 4.22.

Table 4.22 Peroxygen/anesthetic tablet composition (54).

Compound	[%] w/w
Carbamide peroxide	28.00
Pluronic F 127	18.00
Avicel PH200 MCC	16.00
Poly(vinylpyrrolidone)	18.00
Sorbitol (70% solution)	8.55
Sodium saccharin	0.20
Benzocaine	10.00
Flavor	1.25

Furthermore, a bleaching/antimicrobial/antigingivitis mouthwash composition has been described that contains the ingredients shown in Table 4.23.

Several other compositions have been detailed, including a colored stripe gel form, bleaching/antimicrobial/antiplaque mouthwash composition, bleaching/antimicrobial/antigingivitis mouthwash composition, and an oxygenating/antimocrobial/emollient cleansing shower and bathtub spray (54).

The compositions are stable for a long period when stored in a single chamber such as a toothpaste tube. The said emulsified bleaching chemicals become activated only when extruded from the container and exposed to the salivary enzymes (54).

4.3.13 *Antibiotic Toothpaste*

An antibiotic toothpaste has been formulated using an inorganic antibiotic metal-containing composition in an amount that is effective to impart substantial antimicrobial activity within the normal time for brushing the teeth (55).

Commonly, a commercially available toothpaste uses triclosan as an antibiotic agent. Triclosan is believed to quickly kill relevant

Table 4.23 Bleaching/antimicrobial/antigingivitis mouthwash composition (54).

Compound	[%] w/w
Hydrogen peroxide, emulsified	3.00
Carbamide peroxide, emulsified	9.00
Eucalyptol	0.092
Thymol	0.064
Methyl salicylate	0.060
Menthol	0.042
Water, purified	41.93
Alcohol	8.60
Flavor	1.75
Poloxamer	1.00
Sorbitol (70%)	16.75
Sodium saccharin	0.40
Sodium lauryl sulfate	1.00
Benzoic acid	0.35
Sodium benzoate	0.30
Copolymer from vinylpyrrolidone and ethylene oxide	11.00

bacteria during normal tooth brushing and leave a residue which exhibits long-lasting antimicrobial activity.

However, triclosan is an organic compound, and therefore suffers from the disadvantage that an antibiotic resistance can develop over time with continued use. Furthermore, triclosan is suspected of inducing skin irritation (55).

A number of metal ions have been shown to possess antibiotic activity, including silver, copper, zinc, mercury, tin, lead, bismutin, cadmium, chromium and thallium ions. These antibiotic metal ions are believed to exert their effects by disrupting respiration and electron transport systems upon absorption into bacterial or fungal cells. Antimicrobial metal ions of silver, gold, copper and zinc, in particular, are considered safe for *in-vivo* use. Antimicrobial silver ions are particularly useful for *in-vivo* use due to the fact that they are not substantially absorbed into the body.

These ions, however, cannot be practically employed in the current toothpaste formulations because the formulations contain sulfates and phosphates that inactivate the antibiotic activity of these ions.

It has been found that a silver-containing zeolite that is incorporated in a polymer to separate the silver from the phosphates and sulfates, releases silver upon normal brushing at a sufficient rate to quickly impart substantial antimicrobial action to the toothpaste (55). Preferably, a gelling polymer is employed as a barrier layer. Carbopol is an appropriate gelling polymer that is commonly commercially available. Carbopols are crosslinked PAA polymers.

It is possible to employ the barrier layers in several ways. For example, the barrier layer may be employed to microencapsulate individual particles of antimicrobial zeolite. Alternately, several particles of antimicrobial zeolite may be distributed within each *drop* of a barrier material. Assemblages of several antibiotic zeolite particles can be coated with Carbopol.

The coated particles are released from the coating upon brushing, the coatings being of a thickness to allow easy release of the particles, but to protect the silver in the particles from deactivating ingredients. It has been determined that such compositions are capable of exhibiting a long shelf life. It is also possible to microencapsulate individual antibiotic particles, e.g., in starch or agar, and then to incorporate the microencapsulated particles into another barrier layer, e.g., Carbopol. Where the inorganic antibiotic metal-containing composition is a silver salt, the barrier layer can isolate the silver salt from inactivating ingredients (55).

4.3.14 *Compositions with Zinc Ions and Polyphosphate Ions*

Zinc ions exhibit antibacterial effects in the oral cavity. Numerous attempts have been made to prepare dentifrice compositions incorporating zinc ions to take advantage of the therapeutic benefits of reduced plaque, gum inflammation, and gingivitis. However, such formulations are known for their unpleasant taste, often referred to as astringent. Such unpleasant organoleptic experiences often result in reduced compliance to an oral care regimen by the consumer.

Various attempts to disguise or to avoid the unpleasant organoleptic aspects have been made, while retaining or enhancing the therapeutic benefits.

A dentifrice composition has been developed that contains a zinc ion source, a polyphosphate ion source, an anethole, and a silica.

The silica has a mean particle size of about 5–12 μm, an Einlehner hardness (56) of about 1 to about 20, and an oil absorption of about 40 $cm^3(100g)^{-1}$ to less than about 100 $cm^3(100g)^{-1}$.

The composition contains zinc ions and polyphosphate ions in a weight ratio of about 0.1:1 to about 10:1 and has a RDA value of about 100 to about 200 and a pellicle cleaning ratio (PCR) value of about 75 to about 110. The zinc ion source can be zinc citrate.

4.3.14.1 *Radiotracer Dentin Abrasion Value*

Abrasion is a pathological, non-carious tooth loss that most commonly affects the premolars and canines (57). Abrasion frequently presents at the cemento-enamel junction and can be caused by many contributing factors, all with the ability to affect the tooth surface in varying degrees (58).

The sources of abrasion may arise from oral hygiene habits such as toothbrushes, toothpicks, floss, and dental appliance or may arise from other habits such as nail biting, chewing tobacco or another object. Abrasion can also occur from the type of dentifrice being utilized as some have more abrasive qualities such as whitening toothpastes (58).

The RDA value can be determined according to the method recommended by the American Dental Association (59).

The history of the development of the abrasivity limits for dentifrices has been reviewed (60). The RDA scores are summarized in Table 4.24.

Table 4.24 RDA scores (58).

Level	RDA Score
Low abrasive: safe for cementum, dentin and enamel	0 – 70
Medium abrasive: safe for enamel, dangerous for cementum and dentin	70 – 100
High abrasive: dangerous for cementum, dentin and enamel	100 – 150
Very high abrasive: harmful limit, damaging for teeth	150 – 250
Not recommended	250 – >250

The specific steps for determining RDA values are as follows (61):

1. Selection and preparation of teeth:
   Sound, single-rooted permanent teeth that are caries-free and vital at extraction are selected. Teeth are then scraped clean with a scalpel. The crown and root tip of each tooth are removed using an abrasive disc so as to prepare a dentin sample 14 *mm* long and at least 2 *mm* wide at the narrower end. Cut pieces of root (dentin chips) or, alternatively, an additional tooth, are also prepared to be later used in determining a correction factor for self-absorption of radiation.
2. Irradiation of dentin:
   The prepared roots and dentin chips described in Step 1 are exposed to a neutron flux of 2×10^{12} *neutrons* cm^{-2} for 3 *h*.
3. Mounting of the roots:
   After irradiation, the irradiated roots are embedded in a mount of cold-ring dental methacrylate resin and mounted onto a cross-brushing machine. Toothbrushes used throughout the test are 50-Tuft, medium, flat, *Pepsodent* toothbrushes.
4. Preconditioning the dentin surfaces:
   Prior to initial test run, the freshly mounted, irradiated roots are brushed with a reference slurry (10 *g* calcium pyrophosphate plus 50 *ml* of a 0.5% CMC-10% glycerine solution) for 6,000 brush strokes. At the beginning of each subsequent day's test run, the roots are brushed for 1,000 strokes.
5. Test run:
   After preconditioning, the dentin samples are then conditioned with the reference slurry (same slurry as in Step 4) for 1,500 brush strokes at the beginning, during and end of each test run. The test run consists of brushing dentin samples for 1,500 brush strokes with a slurry of test product (25 *g* dentifrice + 40 *ml* deionized distilled water).
6. Preparation of correction factors:
   The correction factors are prepared by dissolving the dentin chips or, alternatively, an additional tooth, from Step 2 in 5 *ml* concentrated HCl brought to a volume of 250 *ml* with distilled water. 1 *ml* of this solution is added to test pastes and reference slurries which are prepared similarly to those in Step 5, and then neutralized with 0.1 *N* NaOH.

7. Radioactive tracer counting:
   The radioactivity of the slurry samples (1.0 *ml*) is determined with an Intertechnique SL-30 liquid scintillation counter. Alternate counting procedure: 3 *ml* aliquotes of each slurry are transferred to stainless steel, flat-bottom 1 *in* × 5/16 *in* planchets and counted using a Nuclear Chicago Geiger Counting System.
8. Calculations:
   The ratio of the average corrected counts for that paste to the average count for the reference multiplied by 100. The reference abrasive is given an arbitrary dentin abrasion value of 100 units.

4.3.14.2 *Pellicle Cleaning Ratio*

The PCR values are determined as measured by the method described in (62,63). This is an *in-vitro* model for the purpose of evaluating the cleaning ability of dentifrices (62). A stained film is slowly deposited on enamel sections and subsequently removed using a mechanical brushing procedure. Results from this test procedure compare favorably with the results of controlled, clinical cleaning studies.

Also, a modified pellicle cleaning ratio test has been used. This test is identical to a previously described method by Stookey *et. al.* (62) with the following modifications (63):

1. A clear artificial pellicle film is applied to bovine chips prior to application of the stained film,
2. Solution heating is used rather than radiative heating during film application,
3. The number of brush strokes is reduced to 200 strokes, and
4. The slurry concentration is 1 part dentifrice to 3 parts water.

In summary, a clear pellicle material is applied to a bovine tooth which is then stained with a combination of the pellicle material and tea, coffee, and $FeCl_3$, which is subsequently treated with the composition, and the change in the reflectance of the tooth surface before and after treatment is the PCR value.

In addition, a modified PCR measurement method was developed (64). The method is characterized by a five-day tea-staining procedure on bovine front teeth slabs on a rotating wheel, standardized brushing of the slabs in a V8 cross-brushing machine, and brightness measurement by a chromametric technique. All tested products were in accordance with the DIN/ISO standard 11609 (65) for toothpastes in terms of dentin abrasivity.

For 41 toothpastes available to European consumers in 1995, the cleaning efficacy was evaluated in comparison with abrasivity on dentin (RDA value) (64). For cleaning power assessment, a modified PCR measurement method was developed. The method is characterized by a five-day tea-staining procedure on bovine front teeth slabs on a rotating wheel, standardized brushing of the slabs in a V8 cross-brushing machine, and brightness measurement by a chromametric technique. All tested products were in accordance with the new DIN/ISO standard 11609 for toothpastes in terms of dentin abrasivity.

Not a single product exceeded an RDA value of 200. The majority of toothpastes (80%) had an RDA value below 100. Only three products surpassed the reference in cleaning power. Most products (73%) had a cleaning power (PCR value) between 20 and 80.

The correlation between cleaning power and dentin abrasion was low ($r = 0.66$), which can be explained by the different influence on dentin and stains by factors like abrasive type, particle surface and size, as well as the chemical influence of other toothpaste ingredients. Some major trends could be shown on the basis of abrasive types. The ratio PCR to RDA was rather good in most silica based toothpastes.

A lower ratio was found in some products containing calcium carbonate or aluminum trihydrate as the only abrasive. The addition of other abrasives, such as polishing alumina, showed improved cleaning power. Some active ingredients, especially sequestrants such as sodium tripolyphosphate or AHBP, also improve the PCR/RDA ratio by stain-dissolving action without being abrasive.

The data for some special anti-stain products did not differ significantly from standard products. Compared with data measured in 1988, a general trend toward reduced abrasivity without loss of cleaning efficacy could be noticed on the European toothpaste market. This may be mostly due to the increased use of high-performance abrasives such as hydrated silica (64).

4.3.15 *Abradable Films*

Abradable films have been developed that are useful in dentifrice formulations (66). Here, the film is stable in the formulation, but abrades to release a pigment or oral care active components upon brushing. The components of such films are shown in Table 4.25.

Table 4.25 Abradable films (66).

Ingredient	[%] per weight
Hydroxypropylmethyl cellulose	1 - 10
Poly(vinyl alcohol)	30 - 60
Polymer matrix	40 - 50
Pigment (Pigment Blue 15)	20 - 60
Plasticizer (propylene glycol)	10 - 30
Surfactant (polysorbate 80)	1 - 5

4.3.16 *Silica Abrasive-Free Composition*

It has been found that dentifrice compositions with an aluminium oxide polishing agent and a water-soluble condensed phosphate salt, such as an alkali metal tripolyphosphate, in the absence of a silica abrasive material exhibit a highly effective cleaning power, excellent gloss and shine whilst at the same time providing low dentine abrasion to the tooth surface (67).

Suitable low calcined aluminas are prepared by mild or weak calcination from aluminium hydroxide. Aluminium hydroxide is converted by calcination into α-aluminium oxide Al_2O_3, which is thermodynamically stable at temperatures above 1200°C. The thermodynamically unstable Al.sub.2O.sub.3 forms occurring at temperatures between 400°C to 1000°C are known as γ-forms (67).

4.3.17 *Toothpaste with Bioadhesive Submicron Emulsion*

Most of the toothpastes with antibacterial components have shown a relatively short antiseptic action due to intensive cleaning of the treated surfaces in the mouth by saliva and a rapid reduction of the concentration of the active component below a minimal inhibiting concentration.

In order to prolong the action of the antiseptics, toothpaste formulation containing antiseptic triclosan and maleic anhydride-vinyl methyl ether copolymer for the improvement of antibacterial action has been developed (68). However, the anionic character of the polymer makes the use of the potent cationic antiseptics in this formulation impossible (69).

Also, the incorporation of chlorhexidine into a toothpaste composition is a problematic task due to the formation of insoluble precipitates with common anions. Some of these compounds are patented (70), however, this approach was unsuccessful due to significant loss of antiseptic activity (69).

A dental composition comprising oil-in-water submicron emulsion for dental use with improved customer properties has been described (71). Although useful, this formulation has a relatively short retention time and rapid elimination from the mucous surfaces at the application.

Sodium lauryl sulfate, needed for submicron emulsion preparation, can cause an irritation in gums and palate and is incompatible with cationic antibacterial substances, e.g., chlorhexidine salts, or alkylammonium derivatives (69).

A toothpaste composition has been described that contains a physiologically acceptable oil, the oil composed of submicron particles, a bioadhesive polymer-coated on the submicron particles, and at least one antibacterial compound (69).

The oil component of the emulsion can be chosen from physiologically acceptable hydrophobic liquids, such as vegetable oils (soya bean, corn, sunflower, coconut, olive, or jojoba), fish or animal oils, synthetic components, such as alkanes, squalane, paraffines, mineral oil, mono- and diol esters (propylene glycol caprylate, isopropyl myristate, dioctyl sebacate, lauroglycol) and glycerides (medium chain triglycerides), polymers, poly(siloxanes), or poly(fluoralcane)s The lipid component is preferably selected from long-chain triglycerides (soya bean oil), medium-chain triglycerides, alcanes (squalane) and esters (isopropyl palmitate, propylene glycol dicaprylate). The concentration of the oil phase can vary most preferably from 5% to 10%.

The antibacterial compound may be chlorhexidine or a chlorhexidine salt, such as bigluconate or diacetate, triclosan, cetylpyridinium chloride, benzalconium chloride and cetyltrimethylammonium bromide. Chlorhexidine in a submicron emulsion demonstrates less staining because of reduced interaction of the oil droplets with tooth dentine and enamel. The bioadhesive coating of the tiny oil particles, charged with chlorhexidine or triclosan leads to a significant prolongation of the drug presence on the mucous surfaces of the mouth, providing an extended release of the antiseptic and flavor components.

The preparation of the submicron emulsion with bioadhesive oil droplets can be achieved using a high pressure homogenizer. At a first stage, a lipid phase is prepared by dissolution of the antibacterial component in the oil phase together with surfactant mixture, antioxidant and flavor component (69).

Subsequently, the oil phase undergoes emulsification in a previously prepared water phase comprises diluted water solution of the bioadhesive polymer, using the usual propeller or rotor-stator mixer. A coarse emulsion is obtained and treated by the high pressure homogenizer in order to obtain submicron emulsion, followed, if necessary, with pH adjusting.

For example, such a composition is prepared as follows (69):

Preparation 4–4: The preparation of 10 *kg* batch of toothpaste is used for the following example. In the first stage, 2.5 *g* of the Methocel E100 M was dispersed in 500 *g* of purified water. This mixture was heated to 90°C and stirred until the polymer was completely hydrated. Then, 2 *kg* of cold water was added and stirred until completely dissolved. Glycerol, in an amount of 700 *g*, was added and stirred until a homogeneous solution was obtained.

At a second stage of preparation, 50 *g* of lecithin and 2 *g* of vitamin E acid succinate and 20 *g* of peppermint oil or *D,L*-menthol were consequently added to 450 *g* of medium chain triglyceride oil (Crodamol TGCC) and stirred slowly until clear solution was prepared. After complete dissolution, 100 *g* of nonionic surfactant Tween-20TM was added and stirred for homogeneous dispersion. This oil phase was combined with 520 *g* of 20% chlorhexidine bigluconate solution followed by intensive stirring.

In a further stage, the product from the second stage described above was mixed with the material from the first stage, using a high speed rotor-stator type mixer, until a homogeneous emulsion was formed. The formed emulsion was treated with high pressure homogenizer at a pressure between 600 L and 1200 *bar*. The emulsion was passed between one to three times through the homogenizer to obtain the desired submicron emulsion. Sodium saccharin in an amount of 20 *g* and 20 *g* of sodium benzoate and 800 *g* sorbitol were added to emulsion and stirred until completely dissolved.

In the final stage, 147.5 *g* hydroxypropyl methyl cellulose was dispersed in 784 *g* of water. This mixture was heated to 90°C and mixed well until all the Methocel E100 M was hydrated. The homogenized submicron emulsion was added to hydrated Methocel at room temperature and mixed until polymer was homogeneously distributed.

Dibasic calcium phosphate was gradually added to formed mixture in an amount of 3.6 *kg*. The mixture was stirred well and then approximately 180 *g* of colloidal silicon dioxide was gradually added until desired viscosity is achieved. The latter number represents a variable amount, this may be adjusted during manufacture to reach desired viscosity of the paste. The viscous mixture was then passed through 60 *mesh* screen, degassed in vacuum to remove entrapped air bubbles and packaged.

Several other special examples of such compositions have also been detailed (69).

4.3.18 Antiplaque Toothpaste Composition

Water-soluble pyridinium compounds, quaternary ammonium compounds and biguanide compounds are effective cationic bactericides for the inhibition of plaque formation; therefore, it has been proposed that these bactericides be blended into toothpaste compositions (18). When these compounds are mixed with anionic ingredients, such as thickening agents or surfactants, in toothpaste compositions, they may cause electric reaction to deteriorate their bactericidal activity because of their cationic properties. To prevent such an electric inactivation of bactericidal activity, a combination of cationic bactericides and nonionic thickening agents or cationic polymers is proposed (72). However, although the bactericidal activity of the cationic bactericides is stabilized in these compositions, these compositions have poor paste-shape retention and poor rheological properties when used (18).

To obtain a toothpaste composition having good shape retention, it has been proposed that cationic polymer be blended into the composition (18). For example, a cationic polymer can be combined with silica to cause flocculation between both substances, so that thickening properties are improved to form a gel.

However, the gel stability is readily deteriorated with time, such as solid-liquid separation, and silica has silanol groups which can absorb any cationic bactericide and thereby inactivate its bacteriocidal activity.

It was found that a certain cationically modified hydroxyethyl cellulose with a particular molecular weight and a particular cationization degree can provide thickening properties without causing any flocculation with silica (18). Therefore, it is possible to obtain a toothpaste composition with stable bactericidal activity and excellent rheological properties by blending such a cationically modified hydroxyethyl cellulose with a water-soluble bactericide selected from pyridinium compounds, quaternary ammonium compounds and biguanide compounds.

The cationically modified hydroxyethyl cellulose should have an average molecular weight of 1 *M Dalton* or higher and a degree of cationization of 0.05 to 0.5 *mol*/glucose. The cationically modified hydroxyethyl cellulose polymers can be obtained by the addition of 3-chloro-2-hydroxypropyltrimethylammonium chloride to hydroxyethyl cellulose (18).

Examples of humectants that can be added are shown in Table 4.26.

Table 4.26 Humectants (18).

Compound	Compound
Glycol	Poly(ethylene glycol)
Xylitol	Poly(propylene glycol)
Lactitol	Maltitol

The wetting agents may be used alone or in combination. The amount of humectant to be used is usually in the range of 5% to 70% by weight, based on the total weight of the composition.

Examples of flavoring agents that can be added are shown in Table 4.27 and in Figure 4.7.

These flavoring agents may be used alone or in combination. The amount of flavoring agent to be used is usually in the range of about 0.1% to 10% by weight, preferably about 0.5% to 5% by weight, based on the total weight of the composition (18).

Examples of sweeteners are saccharin sodium, acesulfame K, stevioside, neo-hesperidyl dihydrochalcone, glycyrrhizin, perillartine, thaumatin, aspartylphenylalanine methyl ester and

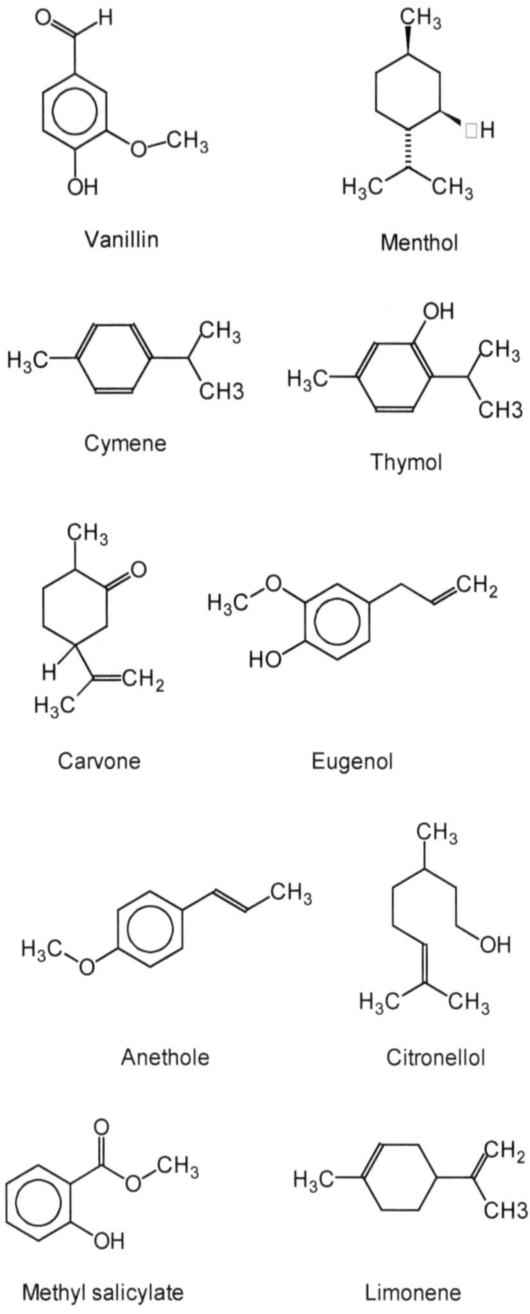

Figure 4.7 Flavoring agents.

Table 4.27 Flavoring agents (18).

Compound	Compound
Menthol	Carvone
Anethole	Eugenol
Methyl salicylate	Limonene
Cymene	*n*-Decyl alcohol
Citronellol	α-Terpineol
Methyl acetate	Citronellyl acetate
Methyl eugenol	Cineole
Linalool	Ethyl linalool
Vanillin	Thymol
Spearmint oil	Peppermint oil
Lemon oil	Orange oil
Sage oil	Rosemary oil
Cinnamon oil	Perilla oil
Gaultheria oil	Clove oil
Eucalyptus oil	

ρ-methoxycinnamic aldehyde. Some of these compounds are shown in Figure 4.3.

Examples of therapeutic agents that can be added are shown in Table 4.28 and in Figures 4.8 and 4.9.

4.3.19 *Saponin*

Natural toothpastes are well known in the oral care industry (73). For example, a toothpaste containing baking soda has been used for many years. A major drawback of prior natural toothpastes has been that they do not perform satisfactorily. That is, a high concentration of baking soda is required to provide adequate cleaning. However, such high concentrations are distasteful and overly abrasive, thus the baking soda toothpastes on the market have insufficient quantities of cleaning agent and high concentrations of flavoring to make them acceptable to consumers.

An organic toothpaste has been developed that contains a mixture of natural ingredients and 0.01–10% by weight of saponin, cf. Figure 4.10, as a surfactant foaming agent (73). The saponin is preferably derived from *Quillaja* or *Yucca* and can be refined to a content of 50–100% from *Quillaja* or 30–100% from *Yucca*.

Table 4.28 Therapeutic agents (18).

Nonionic bactericides	Amphoteric bactericides
Triclosan	Dodecyldiaminoethylglycine
Enzymes	**Fluoro compounds**
Dextranase	Sodium monofluorophosphate
Amylase	Potassium monofluorophosphate
Protease	Sodium fluoride
Mutanase	Stannous fluoride
Lysozyme	
Lytic enzymes	
Other compounds	**Other compounds**
ε-Aminocapric acid	Glycyrrhetic acid
Aluminum chlorohydroxyl allantoin	Glycerophosphate
Dihydrocholesterol	Chlorophyll
Glycyrrhizin salts	Caropeptide

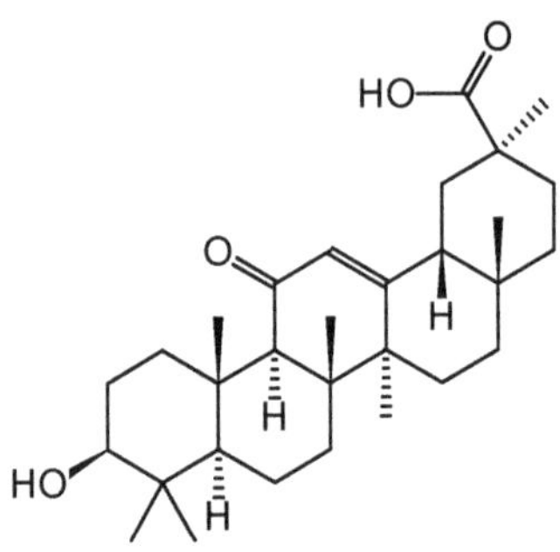

Glycyrrhetic acid

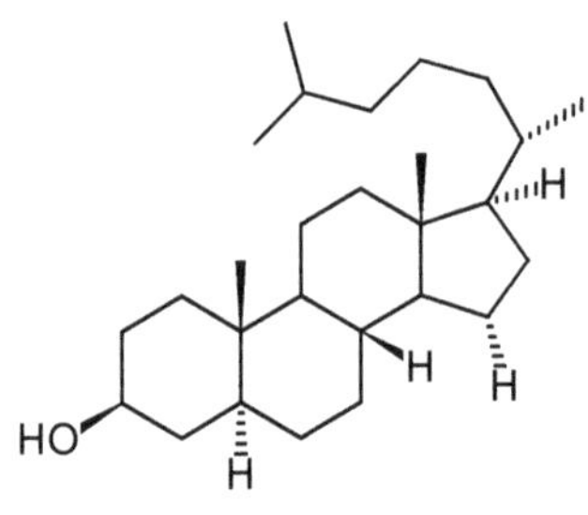

Dihydrocholesterol

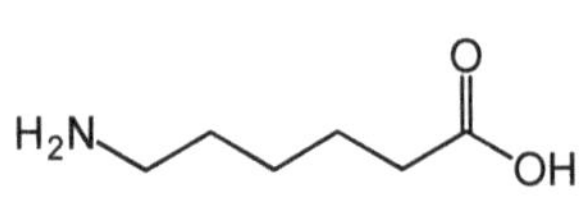

Aminocapric acid

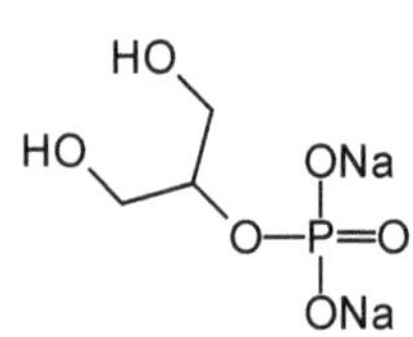

Glycerophosphate

Figure 4.8 Therapeutic agents.

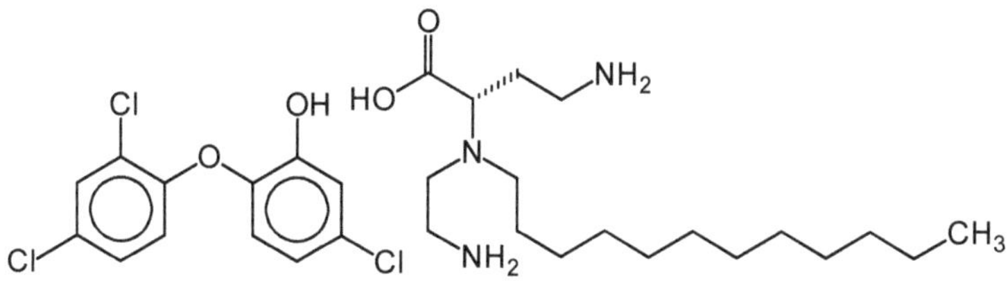

Triclosan Dodecyldiaminoethylglycine

Figure 4.9 Therapeutic agents.

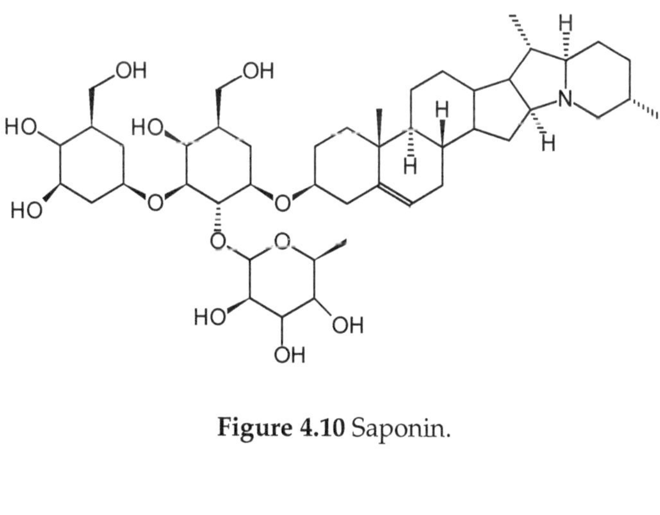

Figure 4.10 Saponin.

Saponins are potential bioactive compounds secreted by plants, endophytic fungi and marine organisms. Saponins are the glycosides containing the non-sugar portion, aglycone (sapogenin), attached to sugar moiety by a glycosidic linkage. The biosynthesis of triterpenoid saponins has been reviewed. Also, the biomedical and pharmaceutical importance of triterpenoid saponins have been detailed (74). The use of saponins as surfactants has been also reviewed in a monograph (75).

The ingredients of the toothpaste are listed in Table 4.29.

Table 4.29 Ingredients of organic toothpaste (73).

Ingredient	Percentage	Purpose
Calcium carbonate	15–50	Whitener
Distilled water	3–18	Moisturizer
Aloe Vera juice	0.5–20	Soother
Fumed silica	0.1–7	Whitener
Hydrated silica	0.5–10	Whitener
Baking Soda	0.5–8	Acid neutralizer
Carrageenan/ Irish Moss	0.1–3	Thickener/soother Irritated tissue
Peppermint oil	0.1–2	Flavor Organic
Manuka oil/ Scoparium oil/	0.5–2	Antimicrobial Native
Grapefruit seed extract	0.1–3	Antimicrobial
Green papaya	0.01–2	Cleanser

The toothpaste containing the above ingredients is unique because it contains therapeutic materials as well as certified organic or wild-harvested ingredients which are tolerant for most individuals. These attributes cooperate with the unique cleansing action of saponin to provide a healthy and effective dentifrice (73).

4.3.20 *Foamable Fluoride Oral Care Composition*

The effective medical management of dental caries is required for populations of patients that exhibit increased risk factors for caries (76).

One practice to reduce dental caries in children is the periodic application, e.g., 1 to 2 times per year, of a foamable fluoride composition having a relatively high concentration of a fluoride-releasing salt, e.g., 1–3% sodium fluoride, that is packaged in an aerosol container in combination with an aerosol propellant.

The composition is dispensed from the container into the trough of a dental tray as a dense, stable, nonflowable foam which is superimposed about and into engagement with the teeth to be treated, to effect fluoride uptake by the dental enamel.

Although conventional dental foams may be effective and are in commercial use, in practice, the thick, dense foam that is produced may cause the patient to experience discomfort during treatment. Additionally, upon completion of treatment, the residual dense foam may be difficult and/or time-consuming to remove from the patient's mouth. For at least these reasons, conventional foams may discourage the usage and the compliance of the patient with the fluoride treatment.

An improved dental fluoride foam composition is needed for the treatment of tooth surfaces that facilitates professional usage and patient compliance so that treatment can be repeated over time to provide effective fluoride treatment for dental caries (76).

A dental fluoride foam composition was developed that contains an aqueous solution of a water-soluble fluoride ion releasable salt, a surfactant selected from the group consisting of nonionic surfactants, zwitterionic surfactants, betaine surfactants and mixtures thereof, and an orally compatible acidifying agent in an amount sufficient to adjust the pH of the composition to about 3 to about 5 (76).

The composition is stable at low temperatures and is a clear solution substantially free of precipitates when held at 4.4°C for 12 *h*. When dispensed from a container into the trough of a dental tray, the composition forms a low density, rapidly collapsible foam which substantially liquefies in about 1 *min* after being dispensed from the foam-generating container and placed in contact with a patient's teeth.

The ingredients of a foamable fluoride treatment composition are listed in Table 4.30 and some compounds are shown in Figure 4.11. The preparation runs as follows:

Preparation 4–5: The formulation is prepared by dissolving the ingredients in water, in the order listed in Table 4.30, at room temperature.

First, 139.5 *g* of the treatment composition and 10.5 *g* of Aeron A-46, a 80/20 mixture of isobutane and propane, are introduced into an aerosol

Table 4.30 Foamable fluoride treatment formulation (76).

Compound	[%] w/w
Deionized Water	90.92
Sodium fluoride	2.65
Sodium benzoate	0.10
Sodium phosphate monobasic	1.38
Sodium saccharin	0.35
Flavor	0.60
Malic Acid	3.00
Cocamidopropyl betaine (30% solution)	1.00

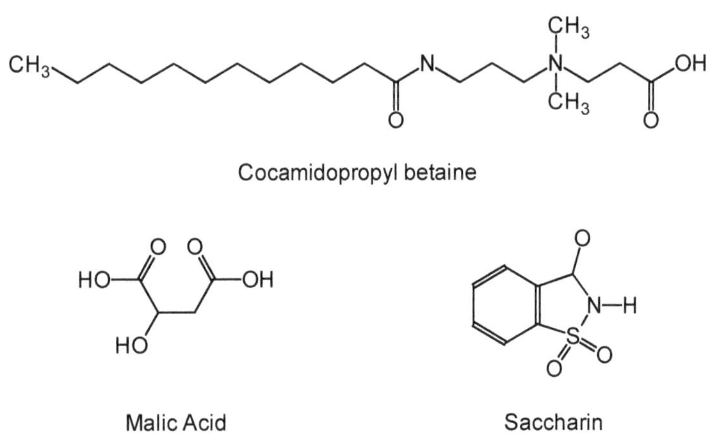

Figure 4.11 Cocamidopropyl betaine, malic acid and saccharin.

container equipped with a valve and nozzle. The composition is then placed in a 4.4°C refrigerator for at least 12 *h*. The resulting composition is clear, not cloudy, and does not have any visible precipitated particles.

A dental foam dispensed from the container collapses within 1 *min* after being dispensed into the trough of a dental tray when placed in contact with a patient's teeth. Upon collapse of the foam, the tray is removed from the patient's mouth and any residual foam is removed by simple water rinsing of the patient's mouth (76).

References

1. Wikipedia, Toothpaste — wikipedia, the free encyclopedia, 2017. [Online; accessed 18-October-2017].
2. T. Stone, *Toothpaste: From Start to Finish*, Blackbirch Press, Woodbridge, Conn., 2001.
3. C. van Loveren, ed., *Toothpastes*, Vol. 23 of *Monographs in Oral Science*, Karger, Basel, 2013.
4. F. Lippert, *An introduction to toothpaste–its purpose, history and ingredients* in C. van Loveren, ed., *Toothpastes*, Vol. 23 of *Monographs in Oral Science*, pp. 1–14. Karger, Basel, 2013.
5. R. Lewis and A.T. Ashcroft, *Tribology of teeth cleaning* in Q.J. Wang and Y.-W. Chung, eds., *Encyclopedia of Tribology*, pp. 3892–3896. Springer US, Boston, MA, 2013.
6. W.W. Harcum and J.J. Modi, Process for making toothpaste using agglomerated dispersible polymers, US Patent 6 258 342, assigned to Hercules Incorporated (Wilmington, DE), July 10, 2001.
7. Frenchbean, The history of teeth cleaning, electronic: https://h2g2.com/edited_entry/A2818686, 2011.
8. I. Van Sertima, *The Golden Age of the Moor*, Transaction Pub., New Jersey, 1992.
9. Wikipedia, List of toothpaste brands — wikipedia, the free encyclopedia, 2017. [Online; accessed 18-October-2017].
10. Wikipedia, Pages in category *Brands of toothpaste* — wikipedia, the free encyclopedia, 2017. [Online; accessed 18-October-2017].
11. B.A. Golden, Non-fluoride containing dietary supplement toothpaste and methods of using the same, US Patent 9 662 294, assigned to Golden Products LLC (Howard Beach, NY), May 30, 2017.
12. S. Chandrasekaran and M. Iyer, Toothpaste comprising calcium based abrasives, US Patent 9 308 158, assigned to Conopco, Inc. (Englewood Cliffs, NJ), April 12, 2016.

13. V.K. Gadkari, V.B. Randive, M.R. Venkat, N.A. Betrabet, and S.A. Kadam, Toothpaste compositions with reduced abrasivity, US Patent 9 072 658, assigned to FMC Corporation (Philadelphia, PA), July 7, 2015.
14. Wikipedia, Chicle — wikipedia, the free encyclopedia, 2017. [Online; accessed 12-November-2017].
15. J.C. Haught, E. Schneiderman, L.A. Sanker, and R.L. Swaine, Jr., Oral care compositions with improved sweetness, US Patent 8 691 190, assigned to The Procter & Gamble Company (Cincinnati, OH), April 8, 2014.
16. N. N., High-intensity sweeteners, Technical report, U.S. Food and Drug Adminsitration, 10903 New Hampshire Avenue, Silver Spring, MD 20993, 2014.
17. W. Contributors, Sugar substitute — wikipedia, the free encyclopedia, 2017. [Online; accessed 16-December-2017].
18. S. Mori and T. Nakajima, Toothpaste composition, US Patent 6 113 887, assigned to Sunstar Kabushiki Kaisha (Osaka, JP), September 5, 2000.
19. L.W. Ritze, Compositions of matter for coloring toothpaste and method of preparing same, US Patent 4 202 878, assigned to The Procter & Gamble Company (Cincinnati, OH), May 13, 1980.
20. J. Uchida, Y. Okimoto, M. Teramae, and M. Mizuno, Mixable and color tone freely reproducible dental coloring material composition, and set and method thereof, US Patent 8 993 650, assigned to Kabushiki Kaisha Shofu (Kyoto, JP), March 31, 2015.
21. S.C. Burgess and A.O. Fatade, Oral care compositions with improved rheology, US Patent 9 572 762, assigned to The Procter & Gamble Company (Cincinnati, OH), February 21, 2017.
22. R. Plata, X. Lu, Y. Zeng, and C. Tan, Processes for preparing toothpaste compositions, US Patent 9 254 253, assigned to Colgate-Palmolive Company (New York, NY), February 9, 2016.
23. B.A. Golden, Non-fluoride containing dietary supplement toothpaste and methods of using the same, US Patent 9 066 889, assigned to Golden Products LLC (Howard Beach, NY), June 30, 2015.
24. C.S. Newby, M.A. Nisbet, and K.S. Price, Alkyl sulfate free and orthophosphate free dentifrice composition comprising fluoride source and a silica dental abrasive, US Patent 9 554 975, assigned to Glaxo Group Limited (Brentford, Middlesex, GB), January 31, 2017.
25. D. Estrada, B.D. Fitz, J.F. Macik, N.J. Napolitano, A.J. Smetana, J.D. Synoids, and M.E. Trama, Dental composition for hypersensitve teeth, US Patent Application 20 030 215 401, November 20, 2003.
26. W.J. Hagar and K.W. Gallis, Continuous silica production process and silica product prepared from same, US Patent 9 327 988, assigned to J.M. Huber Corporation (Atlanta, GA), May 3, 2016.

27. M. Prencipe, A. Morgan, and V.P. Maloney, Oral care composition, US Patent 9 532 932, assigned to Colgate-Palmolive Company (New York, NY), January 3, 2017.
28. S. Mason, A. Levan, R. Crawford, S. Fisher, and A. Gaffar, *Clinical Preventive Dentistry*, Vol. 13, p. 6, January 1991.
29. G. Kowitz, J. Jacobson, Z. Meng, and F. Lucatorto, *Oral Surgery, Oral Medicine, Oral Pathology*, Vol. 70, p. 529, 1990.
30. R. Vaccaro, B. Ghalili, and G.E. Anastassov, Thin film toothpaste strip, US Patent 9 656 102, May 23, 2017.
31. W.A. Mchale and D.G. Brown, Enamel protectant and repair toothpaste, US Patent 9 616 004, assigned to Premier Dental Products Company (Plymouth Meeting, PA), April 11, 2017.
32. W.A. Mchale and D.G. Brown, Enamel protectant and repair toothpaste treatments, US Patent 9 585 818, assigned to Premier Dental Products Company (Plymouth Meeting, PA), March 7, 2017.
33. M. Wong and M. Prencipe, Striped toothpaste stable to color bleeding, US Patent 5 876 701, assigned to Colgate Palmolive Company (New York, NY), March 2, 1999.
34. S.A. Duke, Compositions, US Patent 4 358 437, assigned to Beecham Group Limited (GB2), November 9, 1982.
35. R.E. Stier, J.D. Vidra, and B. Misek, Dentifrices, US Patent 4 568 534, assigned to Beecham Inc. (Clifton, NJ), February 4, 1986.
36. D.K. Kiozpeoplou, Dispensing container of toothpaste which effervesces during toothbrushing, US Patent 4 487 757, assigned to Colgate-Palmolive Company (New York, NY), December 11, 1984.
37. J.E. Grimm, III, Dentifrice containing encapsulated flavoring, US Patent 3 957 964, assigned to Colgate-Palmolive Company (New York, NY), May 18, 1976.
38. J.B. Barth, Flavored dentifrice, US Patent 3 929 988, assigned to Colgate-Palmolive Company (New York, NY), December 30, 1975.
39. J.E. Grimm, III, Dentifrice containing encapsulated flavoring, US Patent 4 071 614, assigned to Colgate Palmolive Company (New York, NY), January 31, 1978.
40. C.M. Kosti, Plaque disclosing dentifrice compositions with semi-solid microcapsules of dye, US Patent 4 348 378, September 7, 1982.
41. N. N., Food and drugs: Listing of color additives subject to certification, Technical Report Chapter 21, Volume 1, Part 74, U.S. Code of Federal Regulations, 732 North Capitol Street, NW, Washington, DC 20401-0001, 2001.
42. H.-U. Scheller, Paste and powder dentifrices, US Patent 4 223 003, assigned to Wurttembergische Parfumerie-Fabrik GmbH (Eislingen, DE), September 16, 1980.

43. R. Plata, X. Lu, Y. Zeng, and C. Tan, Toothpaste composition, US Patent 9 468 592, assigned to Colgate-Palmolive Company (New York, NY), October 18, 2016.
44. G. Szewczyk, N.A. Patel, S. Jogun, and M. Prencipe, Color changing compositions, US Patent 9 522 111, assigned to Colgate-Palmolive Company (New York, NY), December 20, 2016.
45. H.W. Wright, Color changing toothpaste, US Patent 6 419 902, July 16, 2002.
46. T. Boyd, G. Pan, H.M. Trivedi, W. Wang, and N. Lin, Color changing consumer products, US Patent 8 647 648, assigned to Colgate-Palmolive Company (New York, NY), February 11, 2014.
47. L. Fruge, J.R. Brown, S. Herles, K. Panaligan, E.S. Arvanitidou, N. Milanovich, and M. Prencipe, Two phase toothpaste composition, US Patent 8 974 772, assigned to Colgate-Palmolive Company (New York, NY), March 10, 2015.
48. L.M. Kao, Dental cleaning and polishing composition comprising diamond particles, assigned to L. M. Kao (Miami, FL).
49. V.B. Randive and V.K. Gadkari, Toothpaste composition, US Patent 8 524 199, assigned to FMC Corporation (Philadelphia, PA), September 3, 2013.
50. Wikipedia, Carrageenan — wikipedia, the free encyclopedia, 2017. [Online; accessed 29-October-2017].
51. A.D. Ballard, Process for making toothpaste using low levels of carrageenan, US Patent 6 447 755, assigned to FMC Corporation (Philadelphia, PA), September 10, 2002.
52. R. Catiis, R.S. Cabanas, and J. Binshtock, Continuous process for making a non-Newtonian paste or cream like material, US Patent 5 236 696, assigned to Colgate-Palmolive Company (Piscataway, NJ), August 17, 1993.
53. N.F. Stanley, Process for treating a polysaccharide of seaweeds of the gigartinaceae and solieriaceae families, US Patent 3 094 517, assigned to Marine Colloids Inc., June 18, 1963.
54. C.M. Kosti, Bleach stable toothpaste, US Patent 6 759 030, assigned to C. M. Kosti (Mason, OH), July 6, 2004.
55. J.E. Barry and J.A. Trogolo, Antibiotic toothpaste, US Patent 6 123 925, assigned to HealthShield Technologies L.L.C. (West Hartford, CT), September 26, 2000.
56. H. Einlehner, Abrasion testing apparatus, US Patent 4 633 701, January 6, 1987.
57. C. Forbes-Haley, S.B. Jones, M. Davies, and N.X. West, *Dentistry Journal*, Vol. 4, p. 25, 2016.
58. Wikipedia, Abrasion (dental) — wikipedia, the free encyclopedia, 2017. [Online; accessed 11-November-2017].
59. J.J. Hefferren, *Journal of Dental Research*, Vol. 55, p. 563, 1976.

60. S.S. John and D.J. White, *The Journal of Clinical Dentistry*, Vol. 26, p. 51, 2015.
61. M. Prencipe, Dentifrice containing zinc ions and polyphosphate ions, US Patent 9 504 634, assigned to Colgate-Palmolive Company (New York, NY), November 29, 2016.
62. G. Stookey, T. Burkhard, and B. Schemehorn, *Journal of Dental Research*, Vol. 61, p. 1236, 1982.
63. D.E. Rice, Dentifrice compositions, US Patent 5 651 958, assigned to The Procter & Gamble Company (Cincinnati, OH), July 29, 1997.
64. P. Wiilknitz, *Advances in Dental Research*, Vol. 11, p. 576, 1997.
65. Technical Committee: ISO/TC 106/SC 7, Dentistry – dentifrices – requirements, test methods and marking, ISO Standard ISO 11609, International Organization for Standardization, Geneva, Switzerland, 2017.
66. G. Szewczyk, N.A. Patel, and S. Jogun, Abradable films for use in oral care, US Patent 9 498 418, assigned to Colgate-Palmolive Company (New York, NY), November 22, 2016.
67. F. Lippert and R.A. Lucas, Silica abrasive-free dentifrice composition, US Patent 9 314 411, assigned to Glaxo Group Limited (Brentford, Middlesex, GB), April 19, 2016.
68. A. Gaffar, N. Nabi, J. Afflitto, O. Stringer, and M. Prencipe, Antibacterial antiplaque oral composition, US Patent 5 192 531, assigned to Colgate-Palmolive Company (New York, NY), March 9, 1993.
69. J. Schwarz, Toothpaste comprising bioadhesive submicron emulsion for improved delivery of antibacterial and anticaries agents, US Patent 6 117 415, assigned to AlphaRx Inc. (Richmond Hill, CA), September 12, 2000.
70. M. Harrison, Method of making dentifrice compositions containing insolubilized salts of 1,6-di-(*p*-chlorophenyl biguanido) hexane, US Patent 3 937 805, assigned to Colgate-Palmolive Company (New York, NY), February 10, 1976.
71. E. Tabibi and A.A. Siciliano, Oral cavity and dental microemulsion products, US Patent 5 130 122, assigned to MediControl Corporation (Newton, MA), July 14, 1992.
72. A.R. Gallopo, N.I. Ibrahim, and S. Mazzanobile, Dental composition, US Patent 5 176 901, assigned to SmithKline Beecham Corporation (Philadelphia, PA), January 5, 1993.
73. M.J. Olmstead, Organic toothpaste containing saponin, US Patent 6 485 711, November 26, 2002.
74. V.R. Netala, S.B. Ghosh, P. Bobbu, D. Anitha, and V. Tartte, *Int. J. Pharm. Pharm. Sci.*, Vol. 7, p. 24, 2015.
75. W. Oleszek and A. Hamed, *Saponin-based surfactants* in M. Kjellin and I. Johansson, eds., *Surfactants from Renewable Resources*, Wiley Series in

Renewable Resource, chapter 12, pp. 239–250. Wiley, Chichester, U.K, 2010.

76. M.T. Joziak and S.W. Fisher, Foamable fluoride oral care composition, US Patent 8 802 060, assigned to Colgate-Palmolive Company (New York, NY), August 12, 2014.

5

Mouth Rinses

Oral mouth rinse compositions have been used for the prevention of bad breath, elimination of oral microorganisms, and for the whitening of the teeth (1). Oral microorganisms are responsible for bad breath, tooth decay, plaque, and gum diseases such as gingivitis.

5.1 Mouth Rinses with Hydrogen Peroxide

Oral mouth rinses containing hydrogen peroxide have the ability to whiten teeth and to reduce the bacterial flora in the oral cavity (1). Hydrogen peroxide is utilized due to its ability to decompose into water and oxygen. The oxygen then acts as both an antimicrobial agent and a bleaching agent to whiten the teeth.

In the oral cavity, the decomposition of hydrogen peroxide into water and oxygen is aided by the enzyme peroxidase, also known as catalase. One factor that affects the rate of decomposition, is the amount of peroxidase present in the oral cavity. A greater concentration results in a greater decomposition. Thus, a mouth rinse that can cause an increase in the amount of peroxidase exhibits a greater whitening and a greater antimicrobial effectiveness.

A system for cleaning water lines has been presented (2), in particular for dental offices in which the peroxide decomposition is accelerated by the presence of an acidic sulfate. Also, the presence of a disinfectant is needed.

Another factor that will increase the rate of hydrogen peroxide decomposition is the effect of temperature (1). A higher temperature will increase the reaction rate between the peroxidase and peroxide, thus causing a faster onset of whitening and microbial kill. A heated mouth rinse will whiten teeth and kill microbes to a greater extent than an unheated mouth rinse of the same composition.

5.1.1 *Enhanced Activity of Peroxides*

Mouth rinses have been developed that enhance the activity of peroxide (1). This can be achieved in two ways. The presence of orally acceptable inorganic cationic salts of sulfates, bisulfates or pyrosulfates in these rinses enhances the generation of saliva in the oral cavity and thus the level of peroxidase therein. This in turn causes a more rapid and efficient generation of active oxygen from the peroxide which in turn brings about greater whitening and microbial kill.

Also, an *in-situ* formation of sulfate, bisulfate pyrosulfate or mixtures thereof from the reaction between sulfites, bisulfites or metabisulfites and peroxide can be used to form sulfates. Here, a two phase system is used that is mixed just prior to use in the oral cavity.

Part 1 of the system contains an orally acceptable inorganic cation salt of sulfite, bisulfite, metabisulfite or mixture thereof.

Part 2 contains an orally acceptable peroxide in an amount that exceeds the stoichiometric amount required to convert the sulfite, bisulfite, metabisulfite or mixture thereof, by at least 0.5%. Other ingredients, such as flavor or ethanol, may optionally be included (1). An example of such a composition is given in Table 5.1

Table 5.1 Pre-brushing mouth rinse composition (1).

Ingredient	[%] per weight
Ethanol	5.00
Flavor Oil	0.10
Benzoic Acid	0.25
Poloxamer 407(10% Aq. Sol'n)	0.25
Sucralose (1% Aq. Sol'n)	15.00
Water	74.90
Sodium Bisulfate	3.00
Disodium pyrophosphate	1.50
Total	100.00

The preparation of such a composition runs as follows:

Preparation 5–1: Combine ethanol, flavor and benzoic acid. Mix until clear. While mixing, slowly add the Poloxamer, sucralose, and water. Continue mixing and add the sodium bisulfate and disodium pyrophosphate. Mix until a clear solution is obtained.

A two phase whitening mouth rinse *in-situ* formation of sodium bisulfate composition is given in Table 5.2,

Table 5.2 Two phase whitening mouth rinse composition (1).

Ingredient	[%] per weight
Phase A	
Ethanol	10.00
Flavor Oil	0.20
Magnesium lauryl sulfate	0.27
Sucralose (1% Aq. Sol'n)	15.00
Water	61.53
Sodium metabisulfite	4.00
Sodium sulfite	4.00
Tromethamine	5.00
Total Phase A	100.00
Phase B	
Hydrogen Peroxide (50% Aq. Sol'n)	12.00
Water	88.00
Total Phase B	100.00

The preparation of such a composition runs as follows:

Preparation 5–2: Phase A: Combine ethanol flavor oil and magnesium lauryl sulfate. Mix until clear. While mixing, add the sucralose, water, sodium metabisulfite, sodium sulfite and tromethamine. Mix until a clear solution is obtained. Phase B: Combine the hydrogen peroxide and water. Mix until uniform. Combine equal amounts of Phase A and Phase B, just prior to use, to form the final mouth rinse.

Similarly, in place of hydrogen peroxide, urea peroxide or mixtures thereof may be utilized with hydrogen peroxide (1).

5.1.2 *Antiplaque Mouth Rinse*

Dental plaque is a mucous film that harbors bacteria on the teeth (3). Calculus or tartar is an encrustation on the teeth consisting of salivary secretions, food residues and various salts, such as calcium phosphate. Dental plaque, a precursor of calculus, unlike calculus may form on any part of the tooth surface, including the gingival margin.

It is generally acknowledged in dentistry that plaque which contains a combination of pathogenic bacteria is a principal etiological factor associated with periodontal disease, dental infections and caries (3). The usual approach toward combating plaque is by mechanical expedients, such as tooth brushing, dental floss and toothpicks.

However, even the most thorough tooth cleaning fails to eliminate interproximal plaque, yet it is the plaque lodging between adjacent teeth that plays a major role in periodontal disease. It therefore becomes necessary to complement mechanical oral hygiene measures with chemotherapeutic agents to inhibit the development of plaque (3).

Among the many chemotherapeutic agents used for the inhibition of dental plaque are cationic agents such as chlorhexidine. However, there are major drawbacks associated with the use of this agent. For example, it not only has an unpleasant taste, but it also results in staining of the teeth and tongue (3).

Cetylpyridinium chloride is another cationic agent that has been formulated into various mouth rinse products. This antiseptic is usually used in a concentration ranging from 0.05 to 0.10% in mouth rinses. Cetylpyridinium chloride-containing mouth rinses are effective but not as effective against plaque and gingivitis as chlorhexidine, although the use of cetylpyridinium chloride does not result in significant dental staining. Also, triclosan is effective against plaque. Triclosan is an antibacterial agent.

While Triclosan has been shown to be retained in plaque for several hours, the salivary release curve is relatively steep, thereby indicating a rapid release from oral binding sites. On the other hand, when triclosan is incorporated into a copolymer which is retained on oral surfaces, it then provides the oral cavity with a reservoir of triclosan. However, the copolymer makes no contribution to combating plaque (3).

An antiplaque mouth rinse composition has been described that includes two antibacterial agents which when used alone have limited effectiveness, but which when combined coact to promote the delivery to and the retention of these agents in the mouth and soft tissues in the dental region of the oral cavity, thereby enhancing the antibacterial activity inhibiting the development of plaque (3).

One agent, triclosan, is water-insoluble and non-cationic. Therefore, a solubilizer is included in the composition. A preferred solubilizer for this purpose is Tween 20, a poly(sorbate). The other agent, cetylpyridinium chloride, is soluble in water and alcohol and is cationic. When the combination of triclosan and cetylpyridinium chloride is delivered to the dental region, it is adsorbed and retained thereby to afford an enhanced antibacterial activity (3).

In particular, it has been found that the combination of the cationic with the non-cationic antibacterial agents has two significant effects (3):

1. The first being that the combined agents are readily adsorbed in the dental region and retained on the teeth and soft tissue for a relatively prolonged period and are not released by saliva in the oral cavity.
2. The second effect is that these agents coact to afford enhanced antibacterial activity that is highly effective against plaque, particularly interproximal plaque that is not reachable by mechanical expedients.

While quaternary ammonium antimicrobials such as cetylpyridinium chloride have long been used in oral mouth rinses, there is still a need for additional formulations, which provide enhanced antimicrobial activity along with an increased user acceptance (4).

Mouth rinse compositions have been developed that provide increased bioavailable levels of a quaternary ammonium antimicrobial and, thus, improved efficacy (4).

The compositions are formulated to be essentially free of anionic, nonionic or amphoteric surfactants. Surfactants are normally employed to achieve the dispersion of water-insoluble additives such as flavoring oils in oral care compositions. It has been discovered that the presence of such surfactants in compositions containing quaternary ammonium antimicrobials can significantly inhibit their activity.

Specifically, the use of surfactants has been found to decrease the amount of bioavailable antimicrobial and is therefore deleterious to achieving acceptable bactericidal efficacy. The mouth rinse compositions should thus be formulated to be essentially free of such anionic, nonionic or amphoteric surfactants, resulting in an enhanced bactericidal efficacy of the compositions, while being aesthetically pleasing (4).

Preferably, these compositions are essentially free of ethyl alcohol, thereby being suitable for safe use by persons suffering from alcoholism, by pregnant women, and by others who cannot, or should not, use alcohol because of medical and health concerns or, psychological, social and job related reasons. For example, many people cannot tolerate alcohol and must avoid the use of mouthwash compositions containing alcohol. Furthermore, young children, pregnant women, and elderly persons are extremely susceptible to health risks when ingesting large quantities of alcohol. Generally, recovering alcoholics must avoid oral contact with alcohol, as well as persons of certain religious beliefs. Also, those persons afflicted by dry-mouth syndrome or using certain medications, often prefer to avoid alcohol-containing mouthwash compositions since the alcohol tends to remove moisture from the oral tissues and complicate the dry-mouth syndrome or feeling. Formulating with essentially no alcohol may also provide some advantages in the taste of the product by eliminating the *burning* sensation associated with alcohol.

An example of a care mouth rinse formulation can be made by conventional processes by mixing the components shown in Table 5.3.

5.1.3 *Method of Treating Mucositis*

Immunodeficient patients frequently exhibit a condition on the oral mucosa which is clinically described as oral mucositis (6). This condition has no known microbial or viral vector that has been implicated as the causative agent. The immunodeficiency that preceded the appearance of mucositis may arise spontaneously from genetic factors, may be caused by infections, e.g., the HIV virus, or

Table 5.3 Mouth rinse compositions (4).

Ingredient	[%] w/w	[%] w/w	[%] w/w
Glycerin	23.000	23.000	13.000
Cetylpyridinium chloride	0.040	0.065	0.050
Domiphen bromide	–	–	–
Zinc lactate	–	–	0.250
Flavor	0.080	0.160	0.160
Saccharin	0.025	0.025	0.030
Poloxamer 407	–	0.050	0.025
Monosodium Phosphate	0.085	0.053	–
Dibasic sodium phosphate	0.070	0.020	–
Color	0.020	0.020	0.020
Ethanol	–	–	1.200
Water	QS	QS	QS

Here, the term *QS* stands for *Quantum satis*, a Latin term meaning *the amount which is needed* (5).

mucositis may be induced as a result of chemotherapy or radiation therapy for neoplastic diseases. This condition has been difficult to treat and has not responded to treatment with antimicrobial agents.

A method of treatment for mucositis has been discovered, which is based on contacting the diseased sites on the affected area of the mucosa with a combination of triclosan and a cationic antibacterial agent (6). A typical semisolid formulation which is a cream is shown in Table 5.4.

5.1.4 *Formulations for Use with Toothbrush Delivery Device*

Toothbrushes are typically used by applying toothpaste to a bristle section followed by brushing regions of the oral cavity, e.g., the teeth, tongue, and/or gums (7). Tooth brushing is part of a daily oral hygiene activity. Proper dental care involves regular flossing, brushing, rinsing with mouthwash and mouth rinses, and dental checkups.

Dentists generally recommend that an individual brush the teeth for a minimum interval per cleaning, such as two minutes. Despite such recommendations, many individuals, especially young children, do not regularly brush their teeth for the recommended

Table 5.4 Semisolid formulation (6).

Ingredient	[%] w/w
Triclosan	0.1 — 5.3
Cetaryl glucoside	0.5 — 6.7
Cetaryl alcohol	0.5 — 7.7
Coco-Caprylate	0.5 — 6.0
Dicapryl ether	0.25 — 5.0
Sweet almond oil	0.25 — 5.0
Petrolatum	0.5 — 6.0
Dimethicone	0.1 — 5
Phase B CPC	0.01 — 4.4
Glycerin	0.5 — 4.6
Sodium methylparaben	0.01 — 0.03
Sodium benzoate	0.25 — 0.3
Deionized water	10 — 90

minimum interval. Such habits often can be attributed to the individual regarding tooth brushing as a mundane duty with few pleasurable aspects.

Dental checkups to a dentist office are typically the only feedback a person receives on their dental cleaning efforts. There is unfortunately plenty of time in between checkups for poor dental habits to cause problems. For example, gingivitis, periodontal disease, and a host of other problems can be magnified if dental cleaning habits are not rectified promptly (7).

Toothbrush devices with liquid collection systems are known in the art, including those having a capillary fluid system to wick the liquid from the collection area to the bristles. Such devices have been described in several publications (8–12).

Some toothbrush devices have been equipped with fluid reservoirs and systems for delivering auxiliary active agents, such as whitening agents, breath freshening agents, etc. These devices contain various transfer mechanisms that can transport the fluids to the brush or applicator, e.g., a sponge, foams, etc.

Formulations for use with a toothbrush delivery device typically contain 5–45% of an antibacerial agent, 5–70% of an orally acceptable flavorant and a liquid carrier in a needed amount (7).

Special examples of the chemicals that can be used are cetylpyridinium chloride, ethanol, cooling flavor peppermint, flavored tea tree oil, propylene glycol, glycerin, citric acid, sodium citrate, $ZnCl_2$, Poloxomer™, nonionic surfactant, tetrapotassium pyrophosphate, tetrasodium pyrophosphate, sodium sulfate, and sucralose (7).

Laboratory tests were carried out to measure the delivery of such compositions onto a surface that mimics the oral tissue, in particular, porcine tongue. The studies indicated that an average of 45 μl of a concentrated formulation can be dispensed through a wicking system when in contact with soft tissue, when using a toothbrush (13).

A special formulation is shown in Table 5.5.

Table 5.5 Pre-brushing mouth rinse composition (7).

Ingredient	[%] per weight
Glycerin	8
95% Ethanol	5
PEG-40 Sorbitan	40
Diisosterate Flavor	20
Saccharin or Sucralose	2.5
Cetylpyridinium chloride	10
Poloxomer (Pluracare)	0
Propylene glycol	0
Purified water	14.5
Total	100.00

This formulation is substantially equivalent to the standard dose of cetylpyridinium chloride received using Plax Overnight mouth rinse. A dose of only 50 μl of this formulation provided even better reduction in volatile sulfur compounds than brushing with Colgate Total® (7).

5.1.5 *Antibacterial Mouthwash*

In the treatment of periodontal diseases, the elimination of microbial dental plaque is important. In the control of dental plaques, the efficiency of the treatment is increased using antibacterial mouthwashes, as well as mechanical periodontal treatment (14).

Most commonly used mouthwashes are chlorhexidine-containing mouthwashes. Chlorhexidine is an effective antimicrobial agent and it has been used as a topical antiseptic for more than 30 years. However, chlorhexidine has some side effects (14).

It is known that chlorhexidine is toxic at cellular level and it may cause dyeing of teeth and other oral surfaces, defects in tasting, burning tongue, desquamation in the epithelium cells and allergic reactions. Additionally, it may even cause anaphylactic reaction when it is administered orally. Furthermore, it is a chemotherapeutic which increases the formation of tartar in long-term usage.

An antibacterial mouthwash has been developed which does not have any side effects as it is not toxic in the cellular level and can be used for eliminating microbial dental plaques in the treatment of periodontal diseases (14).

This antibacterial mouthwash contains a 0.75% boric acid solution (14). It has been discovered that boric acid (0.75%) is better in clinical periodontal parameters compared to normal saline (control) and chlorhexidine groups when it is used as an adjuvant therapy in chronic periodontitis patients in addition to mechanical periodontal therapy. A three-month follow up of enrolled patients has shown no side effects.

5.1.6 Prevention of Periodontal Diseases

Recent research on the relationship between periodontal diseases and cardiovascular diseases points out that the former diseases possibly induce the latter diseases, such as cardiac attack, cardiac infarction, and brain infarction through intravascular invasions of periodontal disease bacteria and their toxins through the gum to cause intravascular inflammation and induce the formation of thrombus (15,16)

It has also been stated that toxins derived from periodontal disease bacteria may act on the liver and adipocytes, decrease the secretion of insulin to increase the blood sugar level, and result in raising the risk of diabetes-related complications. On the contrary, it has also been revealed that periodontal diseases and diabetes are closely to related each other in that hyperglycemic state, for example, enhances the production of inflammatory cytokines and inflammatory mediators and promotes the disruption of periodontal tissues (16, 17).

It has been found that adenosine *N1*-oxide, cf., Figure 5.1, and its derivatives, including 3′-glucosyladenosine *N1*-oxide, 5′-glucosyladenosine *N1*-oxide, 5′-adenosine diphosphate *N1*-oxide, and 5′-adenosine triphosphate *N1*-oxide, are useful as effective ingredients for a mouthwash composition for preventing and/or treating periodontal diseases (16).

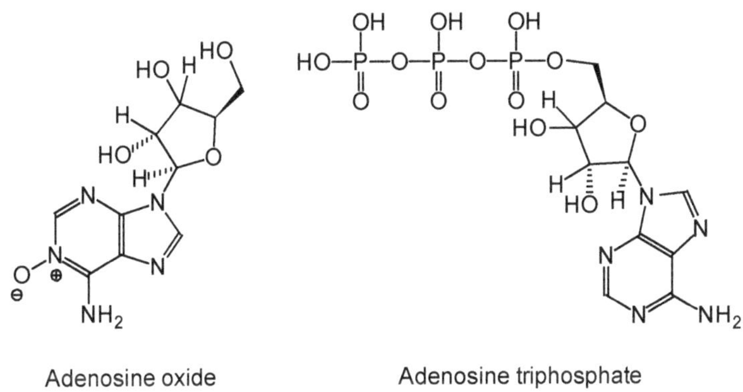

Figure 5.1 Adenosine derivatives.

5.1.7 Alcohol-Free Compositions

Conventional mouthwash compositions frequently contain ethanol, up to 27% volume, as a preservative and antimicrobial agent (18). In some individuals, alcohol may exacerbate xerostomia (dry mouth), which in turn can cause halitosis and significant tooth decay, as the protective effect of saliva's remineralizing the enamel is reduced.

Xerostomia also makes the mucosa and periodontal tissue of the mouth more vulnerable to infection and increases the risk of gingivitis. Mouthwash containing high levels of alcohol also may be undesirable for use by alcoholics, children, and members of certain religious faiths, and some consumers may object to the burning sensation of relatively high alcohol compositions.

Ethanol-free liquid mouthwashes containing cetylpyridinium chloride that do not have an unacceptably bitter taste have been developed (18).

This can be accomplished by admixing relatively insoluble flavoring oils with the cetylpyridinium chloride in a particular ratio.

A liquid oral care composition which is substantially free of ethanol, has been presented, consisting of (18):

1. An antimicrobially effective amount of cetylpyridinium chloride, and
2. Flavoring oils which are substantially insoluble in water at room temperature.

The ratio of the cetylpyridinium chloride to the flavoring oils ranges from 1:1.5 to 1:2.5. It is believed that at optimal concentrations, the cetylpyridinium chloride and the flavoring oils spontaneously or with gentle mixing can form a stable micelle or a water-in-oil microemulsion at room temperature in the presence of water, which helps mask the bitter taste of the cetylpyridinium chloride without affecting its efficacy as an antibacterial agent. At the same time, the surfactant properties of the cetylpyridinium chloride allow elevated levels of poorly soluble oils which themselves may have antibacterial properties (18).

Such a composition is shown in Table 5.6. The flavoring agent in

Table 5.6 Alcohol-free mouth rinse composition (18).

Ingredient	[%] per weight
Sorbitol	3 – 5
Glycerin	6 – 9
Propylene glycol	6 – 8
Sodium fluoride	0.04 – 0.06
Cetyl pridinium chloride	0.05 – 0.1
Sodium saccharin	0.01 – 0.03
Polaxamer 407	0.3 – 0.5
Flavoring agent	0.05 – 0.5
Menthol/coolant	0.02 – 0.06
Dye	0.0 – 0.01,
Parabens	0.02 – 0.06
Water	75 – 85

Table 5.6 contains the components shown in Table 5.7.

Table 5.7 Flavoring agent (18).

Ingredient	[%] per weight
L-Menthol	15 - 25
N-Ethyl-*p*-menthane-3-carboxamide	1 - 3
trans-Anethole	2 - 6
Eugenol	0.1 - 0.3
Peppermint oil	10 - 20
Thymol	0.2 - 0.6
Methyl salicylate	10 - 15
Cinnamic aldehydes	0.1 - 0.3
Spearmint oil	35 - 55

5.1.8 Dual Phase Mouthwash

A preservative system for dual phase mouthwash composition has been developed. This consists of (19):

1. A hydrophilic phase including a hydrotrope,
2. A hydrophobic phase, and
3. A preservative selected from methylisothiazolinone, sodium benzoate, and potassium sorbate and combinations thereof.

A particular dual phase mouthwash composition is listed in Table 5.8.

The composition, when mixed, forms a temporary oil-in-water emulsion which breaks down and separates back into the hydrophobic and hydrophilic phases within 5 *s* to 1 *h* after mixing.

Such compositions are useful to protect the teeth by facilitating repair and remineralization, in particular to reduce or inhibit formation of dental caries, reduce or inhibit demineralization and promote remineralization of the teeth, reduce hypersensitivity of the teeth, and reduce, repair or inhibit early enamel lesions, e.g., as detected by quantitative light-induced fluorescence or electronic caries monitoring (19).

Quantitative light-induced fluorescence has been developed to detect and assess dental caries based on the phenomenon of change of autofluorescence of a tooth by demineralization. However, stains on the tooth surface exhibit the same phenomenon, and therefore quantitative light-induced fluorescence can also be used to measure

Table 5.8 Dual phase composition (19).

Ingredient	Phase A hydrophobic	Phase B hydrophilic
Glycerin	–	7.5
Mineral oil	12	–
Sodium fluoride	–	0.05
Sodium saccharin	–	0.08
Citric acid–anhydrous	--	0.01
Anhydrous monosodium phosphate	–	0.05
Surfactant	–	0.1
Flavor	1.1	–
Dye	0.00012	0.004
Potassium sorbate	–	0.1
Sodium benzoate	–	0.11
Cetylpyridinium chloride	–	0.05
Water	–	Balance
pH	N/A	5.5

the percentage fluorescence change of stained enamel with respect to surrounding unstained enamel (20). Also, the method of electronic caries monitoring has been detailed (21).

In order to optimize the preservative system, different preservatives in Table 5.8 are substituted for potassium sorbate and/or sodium benzoate, and the characteristics of the formulation were tested for their antimicrobial efficacy of the hydrophilic phase, flavor impact, and aesthetics impact.

5.1.9 *Non-alcohol Bioactive Essential Oil Mouth Rinses*

Antiseptic mouth rinses traditionally contain ethanol at fairly high levels, ranging from approximately 20% up to about 30% by volume, based on the total mouthwash volume. Ethanol is used both as a vehicle and as a solvent in which the active ingredients, and additives such as astringents, fluorides, color additives, flavor oils, etc., can be dissolved and then dispersed into solution. Ethanol also provides a preservative role for the mouth rinse during storage and use, and enhances the flavor oil organoleptic cues (22).

However, the use of high levels of alcohol may sometimes be found unacceptable by some mouthwash users. On the other hand, reducing the levels of alcohol in such mouth rinse compositions can have significant disadvantages. Such disadvantages include a reduction in the solubility of the mouth rinse actives and/or the other mouth rinse ingredients.

For example, it has been found that lowering alcohol concentration, i.e., replacing the alcohol with water, in commercially available mouth rinse compositions can result in cloudy or turbid compositions. Cloudy or turbid compositions present a clear disadvantage from an aesthetic point of view since clear mouth rinse solutions are certainly more preferred by consumers than cloudy, turbid or otherwise heterogeneous ones.

Also, lower alcohol concentrations can result in a noticeable decrease in the ability of the composition to kill the oral microorganisms responsible for bad breath, plaque and gum disease. This loss in antimicrobial activity is not only due to the reduction of alcohol as a vehicle, but also to the reduced bioavailability of the solubilized actives.

For these reasons, reduced and/or no alcohol-containing mouth rinses have been developed. In this way, liquid compositions containing oil or oily components which have a reduced turbidity or cloudiness have been presented.

A method of preparation for such formulation consists of (22):

1. Preparing a first premix composition comprised of propylene glycol, an antimicrobial essential oil, and optionally, a flavor,
2. Preparing a second premix composition comprised of propylene glycol, and an organic acid selected from the group consisting of ascorbic acid, sorbic acid, citric acid, glycolic acid, lactic acid and acetic acid, benzoic acid, salicylic acid, phthalic acid, phenolsulfonic acid, succinic acid and mixtures thereof,
3. Preparing a third premix composition comprised of at least one surfactant selected from the group consisting of alkyl sulfates, Poloxamers, and combinations thereof, and an aqueous phase comprised of water, and
4. Adding the first premix to the third premix,
5. Mixing the composition of step 4 until uniform and homogeneous,

6. Adding the second premix to the composition of step 5 and mixing until uniform and homogeneous, and
7. Adding a sugar alcohol solvent to the composition of step 6 and mixing the composition until uniform and homogeneous.

5.2 Properties

5.2.1 Periodontal Disease

Periodontal, disease also known as asgum disease, is a broad term used to describe those diseases which attack the gingiva and the underlying alveolar bone supporting the teeth (4). The disease exists in a number of species of warm-blooded animals, including humans and canines, and includes a series of diseases exhibiting various syndromes which vary from each other according to the stage or situation of the disease or the age of the patient. The term is used for any inflammatory disease, which initially occurs at a marginal gingiva area and may affect the alveolar bone. Periodontal disease affects the periodontium, which is the investing and supporting tissue surrounding a tooth, i.e., the periodontal ligament, the gingiva, and the alveolar bone.

Two common periodontal diseases are gingivitis (inflammation of the gingiva) and periodontitis (inflammation of the periodontal ligament manifested by progressive resorption of alveolar bone, increasing mobility of the teeth, and loss of the teeth at advanced stage). Combinations of inflammatory and degenerative conditions are termed periodontitis complex. Other terms used for various aspects of periodontal disease are juvenile periodontitis, acute necrotizing ulcerative gingivitis, and alveolar pyorrhea.

Periodontal disease may involve one or more of the following conditions (4):

- Inflammation of the gingiva,
- Formation of periodontal pockets,
- Bleeding and/or pus discharge from the periodontal pockets,
- Resorption of alveolar bone,
- Lose teeth, and
- Loss of teeth.

Periodontal disease is generally considered to be caused by bacteria, which are generally present in dental plaque which forms on the surface of the teeth and in the periodontal pocket. Thus, known methods for treating periodontal disease often include the use of antimicrobials and/or anti-inflammatory drugs.

Periodontal disease is a major cause of tooth loss in adult humans. Tooth loss from periodontal disease is a significant problem beginning at age 35, but even by age 15 it is estimated that about 4 out of 5 persons already have gingivitis and 4 out of 10 have periodontitis.

While good oral hygiene, as achieved by brushing the teeth with a cleansing dentifrice, may help reduce the incidence of periodontal disease, it does not necessarily prevent or eliminate its occurrence. This is because microorganisms contribute to both the initiation and progress of periodontal disease.

Thus, in order to prevent or treat periodontal disease, these microorganisms must be suppressed by some means other than simple mechanical scrubbing. Towards this end, there has been a great deal of research aimed at developing therapeutic dentifrices, mouthwashes, and methods of treating periodontal disease, which are effective in suppressing these microorganisms (4).

5.2.2 *Biological Activity of Antimicrobials*

For the evaluation of the bioavailability and the biological activity of quaternary ammonium antimicrobials in mouthwash formulations, an *in-vitro* disk retention assay (DRA) can be used to estimate drug bioavailability, as well as an *ex-vivo* plaque glycolysis and regrowth model (PGRM) to assess the biological activity (4,23).

The DRA is an *in-vitro* method developed to measure the available level of cetylpyridinium chloride in mouthwash formulations. This method is based on the binding of the cationic cetylpyridinium chloride molecule to the anionic surface of a cellulose filter disk (23).

The PGRM is a model used to assess the *in-vivo* therapeutic biological activity of antiplaque and antigingivitis agents with broad spectrum antimicrobial activity that includes generalized actions on glycolysis response of overnight *de novo* plaque biofilms (4). The model is uniquely designed in that it ensures that topical treatment of plaque occurs *in-vivo*, hence plaque is treated *de novo*, as in the clinical situation as an intraoral biofilm. The model permits the sampling of treated plaque samples at timed intervals following rinse exposure, thereby permitting an assessment of the retained activity of antimicrobials post-treatment.

Lastly, the model uses non-treated plaque samples taken from subjects to serve as internal control for treatment comparisons. The method in principle allows for multiple analytical characterizations of *in-situ* antiplaque/antibacterial effects of topical formulations, including assessments of live/dead bacterial populations, regrowth or matrix reproduction capabilities of treated biofilms and metabolic activity of treated biofilms. The glycolysis portion of the test assesses the ability of treated biofilm bacteria to uptake and metabolize dietary sugar to produce acidic end products which are easily assayed either as pH reduction in media buffer or by assessments of the acids produced.

The acid portion of the test offers a convenient, specific and sensitive target for evaluating the formulation activity, which is of primary interest in establishing equivalence of formulation variations, and thus verifying clinical effectiveness. Importantly, the assay has been shown to correlate strongly with the clinical gingivitis and bleeding scores of several products that have similar clinical outcomes (4).

5.2.3 *Effect on the Microhardness*

An *in-vitro* study was designed to evaluate the effect of four mouth rinses on the microhardness of esthetic restorative material (24).

In the study, forty specimens of a resin composite (Filtek™ P60) material (3M ESPE, St. Paul, MN, USA) with 3 *mm* diameter and 3 *mm* height were prepared using a plastic mold. The baseline microhardness values of the specimens were recorded using a Vickers microhardness tester.

The samples were randomly allocated into four groups, each containing ten specimens. The four groups are shown in Table 5.9.

The pH of all the mouth rinses was assessed. Then, the specimens were immersed in 20 *ml* of respective mouth rinses and kept in an incubator at 37°C for 24 *h*. The change in the microhardness values of the specimens was recorded.

Table 5.9 Sample groups (24).

Groups	Manufacturer
Listerine®	Johnson & Johnson
Colgate® Plax	Colgate
Freshclor®	Group Pharmaceuticals
AloeDent®	Optima, Bradford-Italy

The so obtained results were subjected to a statistical analysis using Wilcoxon signed-rank, Kruskal-Wallis, and Mann-Whitney U tests. SPSS software version 16 and Microsoft Excel version 7 were also used. The level of significance was set at P= 0.05.

A reduction in the microhardness of all the specimens irrespective of the mouthwashes was found. The Listerine group showed a significant reduction in the microhardness in comparison to the groups Colgate Plax, Freshclor, and AloeDent.

On the other hand, there was no statistically significant difference between the groups Freshclor, and AloeDent.

All the mouth rinses showed a reduction in the surface hardness of the esthetic restorative material. A high reduction of the surface hardness was shown in the Listerine mouth rinse. There was no statistically significant difference between Freshclor and AloeDent mouth rinses (24).

5.2.4 *Efficacy of Chlorhexidine*

A study was performed to evaluate the incidence of post-surgical adverse events at submerged implant sites as well as the antiplaque, antigingivitis and antistaining effects in the entire dentition of patients that were treated with two mouthwashes (25).

Two 0.12% chlorhexidine-containing mouthwashes were compared for 15 *d*. One with 0.12% chlorhexidine (Composition 1) and 0.1% hyaluronic acid and one with only 0.12% chlorhexidine (Composition 2).

Significant differences were found between the two rinses regarding the presence of oedema within 2 *d* after surgery, i.e., 20% for the patients with composition 1 and 78% for the composition 2. No other significant differences were recorded between the two mouthwashes. No intergroup differences in plaque, staining and gingivitis indexes were registered (25).

5.2.5 *Reduction of Breath Malodor*

Oral malodor is the result of volatile sulfur compounds, carboxylic acids and amines (26). The malodorous compounds are generated primarily through putrifactive action of oral microorganisms on sulfur-containing amino acids, peptones or proteins found in the mouth.

Such microorganisms are readily available in saliva and dental plaque or may be derived from proteinaceous food particles trapped between the teeth, in the gingival crevice or adhering to the mucous membranes and the irregular surface of the tongue, as well as exfoliated oral epithelium, food debris and the like.

Studies have indicated that mouth odor not only comes from the posterior dorsal surface of the tongue but also from periodontal pockets. People with periodontal involvement have an attendant increase in oral malodor from disintegrated epithelial cells (26).

Formulations containing sodium chlorite and a metal ion, such as zinc, capable of complexing with sulfur-containing compounds, can be used for the reduction of oral malodor (26).

5.2.5.1 *Effect of Various Mouth Rinse Products*

The effect of different mouth rinse products on intraoral halitosis has been elucidated (27). Five mouth rinse products where tested that contained zinc acetate and chlorhexidine diacetate, zinc lactate, chlorhexidine and cetylpyridinium chloride, zinc acetate and chlorhexidine diacetate with reduced amounts of mint and menthol, zinc chloride and essential oil, and chlorine dioxide.

It was found that the contents of hydrogen sulfide, methyl mercaptan, and the organoleptic scores were significantly reduced 12 *h* following rinsing with all substances compared to a placebo.

Hydrogen sulfide was more effectively reduced after rinsing with zinc acetate and chlorhexidine diacetate and zinc acetate and chlorhexidine diacetate with reduced amounts of mint and menthol in comparison to rinsing with zinc chloride and essential oil (27).

5.2.6 *Oral Care Composition Containing Ionic Liquids*

An ionic liquid is a class of salt comprising a cation and an anion that is commonly in liquid state below room temperature (28). Ionic liquids generally show a much lower symmetry than conventional salts and the charge of cation and anion is distributed over a larger volume of the molecule by resonance.

Ionic liquids are often composed of a cation comprising a heterocyclic ring and a counter anion, often inorganic in nature. The nature of the cation and anion will determine the hydrophobicity, viscosity, density and other physical parameters and properties of the ionic liquid. Ionic liquids have been evaluated as environmentally friendly or *green* alternatives to conventional organic solvents for a wide range of organic synthetic applications.

Examples of ionic liquids for use in mouth rinses are choline salicylate or tris-(2-hydroxyethyl) methylammonium methylsulfate (28). The composition may further contain diluents, bicarbonate salts, pH modifying agents, surfactants, foam modulators, thickening agents, viscosity modifiers, humectants, sweeteners, flavorants, pigments, antibacterial agents, anticaries agents, anticalculus or tartar control agents, and abrasives.

The ionic liquid is suitable for disrupting and dissolving a biofilm on an oral surface. It has been found that an increase in the concentration of humectants in the composition increases the level of biofilm removal. Glycerin, sorbitol, propylene glycol and mixtures thereof may be used as humectants.

Commercial mouth rinses such as a mouthwash with cetylpyridinium chloride, show no early morning salivary sediment dissolution. However, it was found that formulations with ionic liquids can help to dissolve a salivary sediment (28).

5.2.7 *Composition with Propolis Extract*

It has been reported that propolis contains numerous active compounds, including many classes of polyphenolic compounds, flavones, flavonones, phenolic acid, and esters.

While these compounds are not entirely understood or characterized, they are generally believed to vary based upon the geographical location and where the bees are located. Generally, on a weight basis propolis contains about 45–55% resins and balsams, including, for example, flavonoids, phenolic acids, and esters, about 25–35% waxes and fatty acids, about 10% essential oils, about 5% pollen, including proteins and protein-derived amino acids, and about 5% of other organic compounds and minerals, including trace minerals, vitamins, ketones, lactones, quinones, steroids, benzoic acid and esters, and sugars.

Oral compositions have been developed that contain a propolis extract, a halogenated diphenyl ether nonionic antibacterial agent, an anionic polymeric copolymer of methyl vinyl ether and maleic anhydride, and a source of fluoride ions (29). An exemplary composition is shown in Table 5.10.

Table 5.10 Composition with propolis extract (29).

Ingredient	[%] w/w		
Water	40	—	70
Sorbitol	20	—	30
Hydrated silica	10	—	20
Methyl vinyl ether maleic acid copolymer	1	—	3
Sodium lauryl sulfate	1	—	3
Flavor	0.5	—	3
Titanium dioxide	0.1	—	3
Carrageenan	0.1	—	3
NaOH	0.1	—	3
Sodium salts	0.5	—	3
Triclosan	0.1	—	3
Sweetener	0.1	—	2
Iron oxide hydrate	0.001	—	1
Propylene glycol	0.001	—	1
Sodium fluoride	0.01	—	2
Alcohol	0.001	—	1
Propolis extract	0.0001	—	3
Color	0.0001	—	1

Such oral compositions can be in the form of a mouth rinse, a dentifrice, a confectionary, a medicament, or a film.

References

1. L. Mackles and W.S. Bess, Mouth rinse with enhanced oxygenating activity, US Patent 7 192 573, March 20, 2007.
2. D.F. Day, C.M. Ott, J.A. Mayo, and D. Kim, Multi-component, safe biocidal complex, US Patent 6 692 757, assigned to Board of Supervisors of Louisiana State University and Agricultural and Mechanical College (Baton Rouge, LA), February 17, 2004.
3. B.M. Libin, Antiplaque mouth rinse, US Patent 6 440 395, August 27, 2002.
4. R.L. Roberge and A.L. Wong, Oral care compositions comprising increased bioavailable levels of quaternary ammonium antimicrobials, US Patent 9 241 885, assigned to The Procter & Gamble Company (Cincinnati, OH), January 26, 2016.
5. Wikipedia, Quantum satis — wikipedia, the free encyclopedia, 2017. [Online; accessed 21-November-2017].
6. B. Libin, Method of treating mucositis, US Patent 7 566 740, assigned to BML Pharmaceuticals, Inc. (Chadds Ford, PA), July 28, 2009.
7. S. Mello, M. Patel, S. Kennedy, T. Boyd, E.S. Arvanitidou, J.R. Brown, and S. Chopra, Mouthwash formulations for use with toothbrush delivery device, US Patent 9 271 565, assigned to Colgate-Palmolive Company (New York, NY), March 1, 2016.
8. R. Kaufmann and D.E. Harder, Efficient fluid dispensing utensil, US Patent 6 322 268, assigned to Avery Dennison Corporation (Pasadena, CA), November 27, 2001.
9. R. Kaufmann, Liquid applicator implement, US Patent 6 497 527, assigned to Dataprint R. Kaufmann GmbH (Delmenhorst, DE), December 24, 2002.
10. J.J. Porter, R. Jarvis, and M. Peters, Breath freshener with mouthwash atomizer, US Patent 6 669 390, assigned to J. J. Porter, December 30, 2003.
11. S.P. Dobos, Disposable oral care device, US Patent 7 124 894, October 24, 2006.
12. V.M. Neal, Disposable toothbrush, US Patent 7 311 456, December 25, 2007.
13. J.J. Gatzemeyer, T.J. Boyd, E.J. Jimenez, and B.M. Russell, Oral care implement having fluid delivery system, US Patent 9 167 886, assigned to Colgate-Palmolive Company (New York, NY), October 27, 2015.
14. H. Sema, Antibacterial mouthwash which is nontoxic in cellular level, US Patent 9 616 003, April 11, 2017.
15. L.L. Humphrey, R. Fu, D.I. Buckley, M. Freeman, and M. Helfand, *Journal of General Internal Medicine*, Vol. 23, p. 2079, 2008.

16. K. Kohno, E. Ohashi, and T. Ohta, Mouthwash composition for preventing and/or treating periodontal diseases, US Patent 9 295 627, assigned to Hayashibara Co., Ltd. (Okayama, JP), March 29, 2016.
17. A. Bascones-Martinez, P. Matesanz-Perez, M. Escribano-Bermejo, M.-Á. González-Moles, J. Bascones-Ilundain, and J.-H. Meurman, *Medicina Oral Patologia Oral y Cirugia Bucal*, Vol. 16, p. 722, 2011.
18. R. Vogt, K. Kohrs, E.S. Arvanitidou, and F. Perna, Alcohol-free mouthwash, US Patent 9 192 565, assigned to Colgate-Palmolive Company (New York, NY), November 24, 2015.
19. C. Lewus, G. Szewczyk, S. Mello, K. Smith-Webster, J. Nesta, R. Dillon, E.S. Arvanitidou, and C. Cuiule, Oral care product and methods of use and manufacture thereof, US Patent 9 622 962, assigned to Colgate-Palmolive Company (New York, NY), April 18, 2017.
20. B. Amaechi and S. Higham, Use of quantitative light-induced fluorescence to monitor tooth whitening, in *Proceedings of SPIE - The International Society for Optical Engineering*, Vol. 4249, April 2001.
21. E. Kucukyilmaz, Y. Sener, and M.S. Botsali, *Pediatric Dentistry*, Vol. 37, p. 14E, 2015.
22. C.J. Mordas, D.R. Queiroz, and P.B. Tsai, Methods of preparing non-alcohol bioactive essential oil mouth rinses, US Patent 9 693 944, assigned to Johnson & Johnson Consumer Inc. (Skillman, NJ), July 4, 2017.
23. S.J. Hunter-Rinderle, L.A. Bacca, K.T. McLaughlin, D. Macksood, A.C. Lanzalaco, J. Parran, and M.J. Doyle, *The Journal of Clinical Dentistry*, Vol. 8, p. 107, 1997.
24. R. George and G. Kavyashree, *Journal of International Oral Health*, Vol. 9, p. 55, 2017.
25. A. Genovesi, A. Barone, P. Toti, and U. Covani, *International Journal of Dental Hygiene*, Vol. 15, p. 65, 2017.
26. W.C. Christopfel, Method of reducing oral malodor, US Patent 6 325 997, assigned to Oxyfresh Worldwide, Inc. (Spokane, WA), December 4, 2001.
27. S. Erovic Ademovski, P. Lingström, and S. Renvert, *International Journal of Dental Hygiene*, Vol. 14, p. 117, 2016.
28. M. Patel, G. Picquet, M. Vandeven, M. Hassan, and R. Paredes, Oral care composition containing ionic liquids, US Patent 9 717 667, assigned to Colgate-Palmolive Company (New York, NY), August 1, 2017.
29. N. Fetissova, C. Blanvalet, and P. Lambert, Oral compositions comprising propolis, US Patent 9 005 680, assigned to Colgate-Palmolive Company (New York, NY), April 14, 2015.

6

Toothbrushes and Dental Floss

6.1 Toothbrushes

6.1.1 History

The history of toothbrushes has been detailed (1).

The predecessor of the toothbrush is the chew stick. Chew sticks were twigs with frayed ends used to brush the teeth while the other end was used as a toothpick (2, 3). The earliest chew sticks were discovered in Sumer Mesopotamia in 3500 BC, an Egyptian tomb dating from 3000 BC, and mentioned in Chinese records dating from 1600 BC. The Greeks and Romans used toothpicks to clean their teeth and toothpick-like twigs have been excavated in Qin Dynasty tombs. Chew sticks remained common in Africa and the rural Southern United States. In the Islamic world the use of chewing stick Miswak is considered a pious action and has been prescribed to be used before every prayer five times a day. Miswaks have been used by Muslims since the 7th century (1).

The modern toothbrush originated in China. The first bristle toothbrush resembling the modern one was found in China. Used during the Tang Dynasty (619-907), it consisted of hog bristles (4). The bristles were sourced from hogs living in Siberia and northern China because the colder temperatures provided firmer bristles (5). They were attached to a handle manufactured from bamboo or bone, forming a toothbrush. In 1223, Japanese Zen master Dõgen Kigen recorded on Shõbõgenzõ that he saw monks in China clean their teeth with brushes made of horsetail hairs attached to an oxbone handle. Then, the bristle toothbrush spread to Europe, brought from China to Europe by travelers. It was adopted in Europe during the 17th century.

The earliest identified use of the word toothbrush in England was in the autobiography of Anthony Wood who wrote in 1690 that he had bought a toothbrush from J. Barret.

The European population found the hog bristle toothbrushes imported from China too firm and preferred softer bristle toothbrushes made from horsehair. Mass produced toothbrushes made with horse or boar bristle continued to be imported to England from China until the mid-20th century.

The first mass produced toothbrush was in 1780 in England by William Addis. After saving a small bone from a meal, he drilled small holes into the bone and tied into the bone tufts of bristles, passed the tufts of bristle through the holes in the bone and sealed the holes with glue. The company remained within the family ownership until 1996. Under the name *Wisdom Toothbrushes,* the company now manufactures 70 million toothbrushes per year in the United Kingdom.

Afterwards, by 1840 toothbrushes started to be mass produced in England, France, Germany, and Japan. Pig bristles were used for cheaper toothbrushes and badger hair for the more expensive ones.

Unfortunately, animal bristle was proven not to be an ideal material as it retained bacteria, did not desiccate efficiently and the bristles were often extricated from their intended fixed insertions. In addition to bone, handles were made of wood or ivory.

During the 1900s, celluloid gradually replaced bone handles. Natural animal bristles were also replaced by synthetic fibers, usually nylon, by DuPont in 1938. The first electric toothbrush, the Broxodent, was invented in Switzerland in 1954.

6.1.2 *Types of Toothbrushes*

The types of toothbrushes are collected in Table 6.1. In comparison to a manual brush, a multidirectional electric brush does a better job in reducing gingivitis and plaque. An electric toothbrush performs rotations of its bristles and cleans regions that are hard to reach.

Table 6.1 Types of toothbrushes (1).

Toothbrush type	Toothbrush type
Electric toothbrush	Sulcabrush
End-tuft brush	Chewable toothbrush
Ecological toothbrushes	

Timer and pressure sensors in an electric toothbrush can encourage a more efficient cleaning process. Electric toothbrushes can be classified according to the speed of their movements as: standard power toothbrushes, sonic toothbrushes, or ultrasonic toothbrushes (1).

An interdental or interproximal brush is a small brush, typically disposable, either supplied with a reusable angled plastic handle or an integral handle. This type is used for cleaning between teeth and between the wires of dental braces and the teeth. There is evidence that after tooth brushing with a conventional tooth brush, interdental brushes remove more plaque than dental floss (1). The size of an interdental brush has been standardized in ISO 16409 (6).

A Sulcabrush is a type of toothbrush that can be used specifically for cleaning along the gumline adjacent to the teeth. The bristles are usually shaped in a pointed arrow pattern to allow closer adaptation to the gums. A Sulcabrush is ideal for cleaning specific areas that are difficult to reach, such as areas between crowns, bridgework and crowded teeth. End-tufted brushes may also be used around fixed orthodontic appliances such as braces (1).

A toothbrush with a small round brush head comprises seven tufts of tightly packed soft nylon bristles, trimmed so the bristles in the center can reach deeper into small spaces. The brush handle is ergonomically designed for a firm grip, giving the control and precision necessary to clean where most other cleaning aids cannot reach. These areas include the posterior of the wisdom teeth, orthodontic structures, crowded teeth, and tooth surfaces that are next to missing teeth. This type can be also used to clean areas around implants, bridges, dentures and other appliances (1).

A chewable toothbrush is a miniature plastic molded toothbrush that can be placed inside the mouth. Such chewable toothbrushes are useful for travelers. These types are available in different flavors such as mint or bubblegum and should be disposed of after use.

Other types of disposable toothbrushes include those that contain a small breakable plastic ball of toothpaste on the bristles, which can be used without water (1).

Commonly, toothbrushes are made from plastic materials. However, such brushes may be a source of pollution. In order to reduce the environmental impact, some manufacturers have decided to use biodegradable materials and/or replaceable toothbrush heads. Also, alternative toothbrushes have been developed that consist of wooden handles, often bamboo and bristles made from bamboo viscose or pig bristles.

6.1.3 *Portable Interdental Toothbrush*

In general, an interdental toothbrush has small brush-shaped bristles for cleaning the spaces between the teeth mounted on the end of the handle (7). These devices are oral hygiene instruments used to clean the oral cavity by washing interdental spaces using putting in and pulling out motions of the brush in the interdental space gaps. Use of the interdental toothbrush facilitates the removal of food residue and plaque in hard-to-reach areas of usual toothbrushes, sterilization after dental treatment such as scaling, prevention of plaque after smoking, and cleaning of orthodontic instruments, such as braces, to maintain oral hygiene.

However, there are problems with the cleaning only being conducted using the brush and thus gums become stimulated, bleeding occurs frequently occurs and the ability to remove food residue and plaque is insufficient. Furthermore, if interdental brushes are continuously used, there is an unpleasant smell from the brush due to poor cleanliness, particularly, the growth of bacteria.

A toothbrush has been developed in which a lid containing a cleaning solution is coupled inside the toothbrush with the bristles. Each time when it is used, the toothbrush is immersed in the cleaning solution so that the toothbrush is soaked with the cleaning solution.

However, this method has disadvantages in that the cleaning solution is stored in the lid of the toothbrush and thus the cleaning solution can be discharged outside if the lid is left open due to the user's carelessness.

Also, care should be taken for preventing loss of the cleaning solution stored in the lid when using the toothbrush. Furthermore, it can be troublesome to repeat the motions of putting in and pulling out the toothbrush in the lid every time the user tries to soak the toothbrush in the cleaning solution (7).

These problems can be solved by a portable interdental toothbrush consisting of: A cleaning solution tube filled with a cleaning solution therein, and a cleaning body having a brush assembled to a discharge end of the cleaning solution tube and being wetted by the cleaning solution discharged through the discharge end (7).

Here, the cleaning solution tube is part of a tube body filled with the cleaning solution and a discharge end discharging the cleaning solution which is extended from the tube body; an engaging portion is formed on a neck portion where the tube body and the discharge end meet with each, and a discharge slit for discharging the cleaning solution is formed on a tip of the discharge end, and the cleaning body is configured such that one end thereof is inserted into the discharge end to form an engaging protrusion for engaging with the engaging portion and, in the other end, an adhesion protrusion liar adhering closely to a discharge slit of the discharge end is formed in the inside surface and a communicating hole for communicating with the brush is formed on a tip (7).

This type of portable interdental toothbrush can be carried conveniently since the cleaning solution tube stays in a sealed state when carried without using it, which makes it convenient to use. When using the interdental toothbrush, an engaging protrusion fixed to an engaging portion is released to wet a brush with the cleaning solution filled in the cleaning solution tube through the discharge hole by pulling the brush body to use the interdental toothbrush; therefore the structure is simple, the production is easy, and production cost is low, thereby allowing any user to use it, which promotes dental and oral health (7).

6.1.4 *Tapered Bristle Filaments*

A toothbrush has been developed that incorporates tapered bristle filaments in combination with tuft holes of a reduced size, such as between 1.3–1.4 *mm* in diameter (8).

The toothbrush balances the number of tuft holes and the number of tapered bristle filaments within each tuft hole to achieve a unique ratio that reduces the cost of manufacture of the toothbrush while maintaining a mouthfeel that consumers associate with a quality product (8).

6.1.5 *Free-Moving Flexible Bristles*

A common toothbrush is designed such that bristles forming a brush are implanted at regular intervals on a head having the shape of a flat plate (9). In order to maximize the brushing effect by causing the brush to come into close contact with a curved surface of the teeth, a variety of methods are used. Different lengths may be assigned to the bristles to adjust the height thereof, large or small diameters may be provided to the bristles, or the bristles may be arranged in various patterns.

However, a conventional toothbrush is problematic in that bristles are implanted on a head having the shape of a fixed flat plate. This has the result that the bristles cannot fully come into contact with the teeth during brushing, a contact area is small, a frictional force is low, and thereby a frictional force of the bristles does not act uniformly on the surface of the teeth during the act of brushing. This results in the brushing effect inevitably being lowered.

Toothbrushes have been developed in which a bristle body implanted with bristles is configured to flexibly move upward, downward, leftward and rightward in a head of the toothbrush according to the dental shape, thus increasing the contact area as well as the frictional force between the teeth and the bristles and thereby maximizing the removal of dental plaque and food remnants by brushing (9).

6.1.6 *Full-Text Contour-Adjustable Toothbrush*

A contour-adjustable toothbrush is a device that provides a more thorough brushing experience while eliminating the need to remove the device from the mouth during brushing (10).

A first brushing assembly and a second brushing assembly are slidably engaged into a housing handle that may be grasped by the user. The first brushing assembly and the second brushing assembly each include a flexible brush head with a concave surface. The concave surface of the first brushing assembly is oriented away from the concave surface of the second brushing assembly, enabling the first brushing assembly and the second brushing assembly to clean the upper teeth and lower teeth simultaneously.

In addition, the first brushing assembly and the second brushing assembly are independently adjustable in order to adjust the length of the device. An electronic variant of the device includes a power supply and at least one actuator (10).

6.1.7 *Poly(trimethylene terephthalate) Bristles*

Toothbrushes typically consist of a head with a handle, with the head having a number of tufts which are used to do the actual cleaning. Typically the head is comprised of a number of bristle tufts which are arranged in an appropriate configuration. The bristle tufts are actually made up of a number of individual bristles which can be anchored into the brush head in any appropriate manner.

Two properties dictate whether a polymer is suitable as a toothbrush bristle component (11). First the flexural stiffness and second the flexural recoverability. These properties also dictate the size of the bristles, e.g., a bristle comprised of a polymer with high flexural stiffness and recoverability will be stiffer than one with a low flexural stiffness and recoverability, thus allowing the bristles to be thinner, allowing more bristles to be packed together, thus providing a greater surface contact area.

The flexural recoverability correlates to the tensile recoverability, which is a standard industrial statistic for a material. A material with a high tensile recoverability will be able to resist splaying when used as a toothbrush bristle material. Splay is the permanent bristle deformation that results from the cyclical flexural strains induced during the tooth brushing process. Tensile recoverability and, therefore, splay resistance is determined both by the type of polymer and how it is processed.

A toothbrush has been described in which the filaments of the brush are made from poly(trimethylene terephthalate) (PTT) (11).

It has been found that PTT exhibits a superior flexural recoverability while having a similar axial elasticity modulus to nylon 6,12. We have also found that some polymeric materials with a similar structure to PTT, e.g., poly(ethylene terephthalate) (PET) and poly(butylene terephthalate), have significantly poorer flexural recoverability than PTT (11).

PTT can be coextruded with other polymers, for example, polymers which have a high flexural elastic modulus. An example of such a coextrusion polymer is PET, which can be made with a higher flexural elastic modulus (10 *G Pa*) than other polymers, such as PBT (3 GPa). Of the possible coextrusions, a preferred embodiment is that filaments are coextruded with a PET core and a PTT sheath, with coextrudates generally offering a balance between cleaning efficiency and splay resistance to be optimized for a given toothbrush (11).

6.2 Dental Floss

The use of dental floss and other interdental cleaners is an important part of dental hygiene, and are used to remove plaque and other particulate from between the teeth and under the gumline, e.g., areas in the mouth where a toothbrush cannot reach (12). Frequently, these are the initiation sites of tooth decay, especially if not cleaned regularly. However, even with routine maintenance, caries and gingivitis still develop in these areas.

Dental flosses are generally linear strips of a material having a fixed diameter and fixed dimensions. However, teeth are not all equally spaced apart. Thus, use of a dental floss that has a diameter less than the distance between the teeth results in inefficient or ineffective cleaning between the teeth. In addition, a space usually resides between the gum and two adjacent teeth that is usually larger than the diameter of dental floss, and efficient cleaning of such an area is difficult. Dental floss users occasionally use dental floss to massage the gums, but dental flosses generally are hard when pulled taut, resulting in potential damage to the gum (12).

6.2.1 *History*

The history of dental floss is somewhat in question, but according to most sources, credit for the invention of dental floss goes to a New Orleans dentist, who in 1815 began advising his patients to use a thin silk thread to clean between their teeth (presumably the battle of New Orleans, also in 1815, garnered more press in the city at the time) (13). Regardless, eventually the idea caught on, and in 1882 a company called the Codman and Shurtleft Company, based in

Randolph, Massachusetts, began marketing an unwaxed silk dental floss. This was followed in 1896 by the first dental floss from Johnson & Johnson. The New Jersey-based J&J took out a patent for dental floss in 1898 that was made from the same silk material used by doctors for silk stitches.

During the 1940s, nylon replaced silk as the material for dental floss; its consistent texture and resistance to shredding were an improvement over the silk versions. The use of nylon also allowed for the development of waxed floss in the 1940s, and for the development of dental tape in the 1950s. Since then, the variety of types of dental floss has expanded to include newer materials such as Gore-Tex, and different textures such as spongy floss and soft floss. And today's floss has other features to make flossing easier. For example, floss with stiffened ends is designed to help with flossing around braces or other dental appliances. Today, floss is a key part of the recommended twice daily brushing, daily flossing, and regular use of a mouthwash that make up a healthy routine for maintaining one's oral health (13).

6.2.2 *Poly(amide) Poly(ethylene oxide) Composites*

Continuous fibrous composite tapes of poly(amide) 6 (PA 6)/ poly-(ethylene oxide) (PEO) containing more than 16,000 individual ribbon-like PA 6 micro-/nano-fiber domains could be produced using a coextrusion and two-dimensional multiplication technique (14).

This melt-based process is suitable for upscaling and applicable to any melt-processable polymers, and highly adjustable in terms of material composition, tape geometry, and number/size of the fiber domains.

The PA 6 fibrous tapes were post-oriented to achieve superior mechanical properties that are comparable with commercial strong polymeric fibrous composite tapes. The high tunability of tape and fiber structure and properties make the coextruded PA 6/PEO tapes a desirable candidate for various strong tape applications (14).

6.2.3 *Elastomeric Dental Floss*

Dental floss compositions have been described with an elastomeric matrix containing one or more particles, and methods of making and using the same (12).

It has been discovered that a gel composition may be effectively used as a dental floss. The elastomeric dental floss consists of at least an elastomeric block copolymer from styrene ethylene butylene styrene, poly(propylene) (PP), a plasticizer, one or more abrasive particles, an optional flavorant, and an optional colorant.

Plasticizers that can be used, include rubber processing oils, such as paraffinic and naphthenic petroleum oils, highly refined aroma-free paraffinic and naphthenic food and technical grade white petroleum mineral oils, isopropyl myristate, and synthetic liquid oligomers of polybutene, polypropene, polyterpene, and others.

Flavoring agents that may be used include essential oils as well as various flavoring aldehydes, esters, alcohols, and similar materials. Examples of the essential oils include oils of spearmint, peppermint, wintergreen, sassafras, clove, sage, eucalyptus, marjoram, cinnamon, lemon, lime, grapefruit, and orange. Also useful are menthol, carvone, and anethole.

Suitable abrasive particles are silica, precipitated calcium carbonate, mica, crystalline aluminosilicate, precipitated silica, glass beads, glass bubbles, glass microspheres, ceramic microspheres and high melting point polymers with a melting point greater than 200°C.

For the fabrication of the elastomeric dental floss, the PP and the elastomeric copolymer preferably are blended and melted together. The plasticizer is preferably added in an amount sufficient to cause the gel to retain a solid structure, but also be deformable.

The mixture then may be extruded or injection molded into sheets or strands and allowed to cool. The extruded sheets and strands may be of any diameter. Preferably, the diameter of the extrusion is such that when the gel is stretched 150% to 1,000% the narrowed diameter of the gel composition may easily slide between the teeth.

Furthermore, the elastomeric dental floss preferably can be elongated up to 1,500% of its initial length without breaking. It has a tensile strength peak at breaking point, for a 2 *mm* wide sample of less than 20 *N min*, and has an elongation at breaking point, for a 2 *mm* wide sample, of more than 100% (12).

6.2.4 *Bristled Dental Floss*

A bristled dental floss has been developed for effectively removing plaque, tartar, food particles and the like from between the teeth of the user (15). The bristled dental floss includes an elongated, flexible floss core. A plurality of floss bristles extends from the floss core. At least one elongated bristle group is formed by the plurality of floss bristles. An undulating pattern is formed by the plurality of floss bristles in each bristle group.

In typical use, the bristled dental floss may be inserted between the teeth of a user and moved in a back-and-forth motion. The bristle groups effectively remove plaque, tartar, food particles from between the user's teeth, particularly in hard-to-reach recessed areas between the teeth (15).

6.2.5 *Star-Shaped Microfiber Dental Floss*

A dental flossing apparatus has been described that contains one or more star-shaped microfiber strands with two nodules and one or more optional therapeutic agents. Also, a method for using the dental flossing apparatus to deliver the therapeutic agents to the teeth and gums has been shown (16).

Unlike a conventional dental floss, the star-shaped microfiber strands absorb and hold about seven times their weight in liquid, much more than natural fibers such as cotton. Thus, the dental flossing apparatus can administer much more therapeutic agent than natural fibers or the typical non-absorbent and sometimes waxed dental floss. Furthermore, the consumer can choose from any number of therapeutic agents and can control the amount of the agent that is applied directly between the teeth and gums.

This dental flossing apparatus has the following advantages (16):

1. The nodules are integrally manufactured along the strand of floss to facilitate handling. These nodules eliminate the waste and discomfort associated with having to wrap the floss around one's fingertips. Approximately 50% of regular dental floss is discarded without ever touching a tooth because it is wrapped around a finger in order to make the other 50% manageable. The nodules also eliminate the need for devices that have been invented to avoid the necessity

of wrapping the floss around a finger, for example, rings to which the floss must be tied and plastic yokes. The plastic yokes resemble a wishbone across the gap of which the floss is either tied or manufactured in place.

2. The star-shaped microfiber increases the effectiveness of the floss significantly. The star-shaped microfiber floss does not just dislodge food particles and plaque, leaving them in one's mouth, as regular flosses do. Rather, the star-shaped microfiber works on a microscopic level to actually glom onto, not only food particles and plaque, but also bacteria, and to remove them from the mouth to be discarded with the floss.
3. The multi-strand design of star-shaped microfiber floss makes the microfiber's inherent proclivity to glom onto microscopic particles, for example, bacteria, viruses and bits of food, even more effectively by allowing the individual strands to work together by attaching themselves to a given particle from more than one side, as a single strand of star-shaped microfiber would do. This multi-strand design also facilitates the positioning of the floss between the teeth in two ways. If the teeth are tightly packed, the several strands of microfiber tend to align themselves one after the other in single file, which allows them to slip more easily between the teeth than one thicker strand would. If, however, the teeth are too tightly packed, the user may twist the several strands into one tighter, stronger strand, as needed, to force the star-shaped microfiber floss between the tightly packed teeth without breaking the floss. In both scenarios, once between the teeth the star-shaped microfiber floss can relax and expand back to its normal multi-strand configuration to be most effective in cleaning.
4. The star-shaped microfiber floss absorbs about seven times its own weight in liquid, much more than cotton, and regular waxed or unwaxed dental floss, allowing the star-shaped microfiber floss to be used to administer therapeutic agents in far greater quantities directly between the teeth and gums.
5. The star-shaped microfiber floss comes untreated with any therapeutic agent, but is ready to absorb and transfer whatever therapeutic agent the consumer may desire directly to the area between the teeth and gums. This is accomplished

by dipping the star-shaped microfiber floss in the desired agent and flossing normally.

6.2.6 *Dental Floss Inside a Dental Tubule*

To maintain dental hygiene, it is important to clean the teeth often, and to remove food remnants from between the teeth. Various devices can be used for this purpose; dental floss is one of the most efficient means currently available. One particularly popular means is lengths of waxed dental floss, contained within a dispenser, wound onto a spool, with one end protruding slightly. In preparation for use, the user pulls the end of the dental floss, exposing the desired length of floss from the dispenser, and then cuts it and uses it.

The length of the dental floss contained within the dispenser can be 50 *m*. A typical length of the length cut for the purpose of use can be approximately 30 *cm*, although the American Dental Association recommends flossing at least once a day, using a dental floss having a length of about 46 *cm* (17).

A method and a dental floss device for flossing the teeth of a user, and a production process of the dental floss device has been presented (17).

The dental floss device includes a single-use dental floss packaged inside a dental tubule. The dental tubule does not practically limit the bending capability of the dental floss wherein the dental floss device is adapted to be inserted into a ball, which takes up at least 70% of the ball internal volume (17).

6.2.7 *Surface Treated Dental Floss*

A number of approaches have been suggested to improve the biocompatibility and the blood compatibility of medical devices (18). One approach has been to modify the surface of the material to prevent undesirable protein adhesion by providing the material with a low polarity surface, a negatively charged surface, or a surface coated with biological materials, such as enzymes, endothelial cells, and proteins. Another approach has been to bind anticoagulants to the surface of biologically inert materials to impart antithrombogenic characteristics to the materials. Still another approach has been the copolymerization of various phospholipids which are used as coating materials for various substrates. Partial polymeric backbone coatings have also been used in a similar fashion. However, many of these methods can result in a leaching or *stripping off* of the coating.

A number of materials have been proposed to be coupled to conventional dental floss. For example, a poly(tetrafluoroethylene)-based floss soaked in chlorhexidine has been proposed, while in another floss the material has been dipped in silver nitrate. Breath fresheners, anticavity ingredients, and a variety of antiseptics have been proposed for coupling to a dental floss, but no proposal has yielded a convenient method of coupling the desired coating.

A surface treated dental floss has been described (18). The methods for preparation include an acrylic surface treatment for coupling thereto of a heparin surface treatment, a collagen surface treatment or both, and an amino-functional polysiloxane surface treatment for coupling thereto of a heparin surface treatment. Such a surface treatment allows a secure method of cost-effectively coupling any of these desired materials as a bonded surface treatment to the floss.

Although any desired biomolecule surface coating may be bonded to the floss, poly(hexanide), i.e., poly(hexamethylene biguanide) is deemed to be a first choice. This is a polymer that is used as a disinfectant and is antiseptic (18).

6.2.8 *Relationship Between the Use of Dental Floss and the Development of Peri-implantitis*

A possible relationship between the use of dental floss or superfloss and the development of peri-implantitis has been assessed (19).

Ten patients with progressive peri-implantitis and with an intensive oral hygiene protocol, which consisted of extensive professional supramucosal and submucosal cleaning, and who were not responding to this therapy, were scheduled for examination. Plaque and bleeding indices and probing depth measurements were assessed, and a radiographic examination was performed every two years.

In all ten cases, remnants of dental floss were found around the neck and the coronal part of a dental implant. After careful removal of these floss remnants and implant cleansing, a significant improvement in the peri-implant conditions in nine of ten cases was noted.

In one case the peri-implant probing depth decreased substantially. However, bleeding on probing was still present.

In-vitro tests showed that the application of various types of dental floss on the exposed rough implant surfaces may easily lead to tearing of floss fibers. This may result in the deposition of floss remnants on rough implant surfaces, which, in turn, may lead to the development of plaque-related peri-implant inflammation and, subsequently, bone loss (19).

6.2.9 *Efficacy of an Unwaxed Dental Floss Impregnated with 2% Chlorhexidine*

The antiplaque and antigingivitis efficacy of a dental floss impregnated with 2% chlorhexidine digluconate was assessed (20).

After 15 *d* the test group had the lowest mean plaque index of 1.04±0.67, thus showing a statistically significant difference compared to the negative control group, in which no interproximal cleaning was performed (1.40±0.65). The mean reduction was 70.2% and 87.26%, respectively. So, no statistically significant difference was found between both of the test groups (20).

6.2.10 *Mechanical and Physical Properties of Various Types of Dental Floss*

To compare maximum load, percentage of elongation, physical characteristics of 4 types of dental floss have been assessed (21):

1. Thai Silk Floss (silk, waxed),
2. Oral B® Essential Floss (nylon, waxed),
3. Experimental Floss Xu (nylon, unwaxed), and
4. Experimental Floss Xw (nylon, waxed).

These types of floss were tested with a universal testing machine (Instron®). Each sample (30 *cm* long, 5 *cm* segment) was fixed and pulled apart with load cell of 100 *N* and a test speed of 100 $mm\ min^{-1}$.

The physical characteristics were investigated with a digital microscope under 2.5±10 magnification, and scanning electron microscope under 1±100 and 5±100 magnification. The size of the filaments was measured in microns (μm) and the fineness was measured in deniers.

For the mechanical properties, the maximum load and the percentage of elongation were presented as mean standard deviation. The distribution of the data was calculated by the Kolmogorov-Smirnov test. One-way ANOVA and multiple comparison (Tukey HSD) were used to analyze the differences among the groups with the level of a statistical difference at $p < 0.05$.

Tukeys HSD (honest significant difference) test is a single-step multiple comparison procedure and statistical test (22, 23). It can be used on raw data or in conjunction with an ANOVA (analysis of variance) to find means that are significantly different from each other. The test is named after John Tukey.

Analysis of variance (ANOVA) is a collection of statistical models used to analyze the differences among group means and their associated procedures, such as variation among and between groups (24). This method was developed by statistician and evolutionary biologist Ronald Fisher in 1921 (25).

The maximum load of Floss Xu, Floss Xw, Oral B and Thai Silk were 47.39, 46.46, 25.38, and 23.70 N, respectively. The percentage of elongation of Oral B, Floss Xw, Floss Xu and Thai Silk were 72.43, 44.62, 31.25, and 16.44%, respectively.

All four types of dental floss showed statistical differences in both the maximum load and percentage of elongation at $p < 0.05$, except for maximum load between Floss Xw and Floss Xu that showed no statistically significant difference.

The physical characteristics of the Thai silk revealed the most disintegrated, the smallest, and the least fine filaments. Floss Xu had the highest maximum load. Oral B had the highest percentage of elongation. Wax coating on Floss X increased the elongation but had no significant effect on the maximum load. The physical characteristics of Thai Silk resulted in the lowest mechanical property values (21).

References

1. Wikipedia, Toothbrush — wikipedia, the free encyclopedia, 2017. [Online; accessed 25-November-2017].
2. C. Panati, *In the Bathroom*, chapter 8, pp. 208–209. Harper Collins, 2013.

3. Z.-R. Zhou, *Introduction to teeth* in Z.-R. Zhou, H.-Y. Yu, L.-M. Qian, and J. Zheng, eds., *Dental Biotribology*, chapter 1, pp. 18–19. Springer, 2013.
4. J.V. Kumar, *Oral hygiene aids* in S.S. Hiremath, ed., *Textbook of Preventive and Community Dentistry*, chapter 38, pp. 412–413. Elsevier India, New Delhi, 2nd edition, 2011.
5. N.O. Harris and F. García-Godoy, eds., *Primary Preventive Dentistry*, Appleton & Lange, Stamford, Conn., 5th edition, 1999.
6. Technical Committee: ISO/TC 106/SC 7, Dentistry – oral care products – manual interdental brushes, ISO Standard ISO 16409, International Organization for Standardization, Geneva, Switzerland, 2016.
7. S.G. Lee, Portable interdental toothbrush, US Patent 9 814 301, assigned to Sang Geun Lee, November 14, 2017.
8. Y.M. Ji, Toothbrush, US Patent 9 788 643, assigned to Colgate-Palmolive Company (New York, NY), October 17, 2017.
9. P.H. Lee, Toothbrush, US Patent 9 788 642, assigned to P. H. Lee, October 17, 2017.
10. S. Jackson, Contour-adjustable toothbrush, US Patent 9 776 317, October 3, 2017.
11. D.G. Savill, Toothbrush and method of cleaning teeth, US Patent 6 053 734, assigned to Chesebrough-Pond's USA Co. (Greenwich, CT), April 25, 2000.
12. C.S. Wong, J.E. Fontana, and P. Focassio, Elastomeric dental floss, US Patent 8 967 161, assigned to Colgate-Palmolive Company (New York, NY), March 3, 2015.
13. J.a. Stofko, Dmd, Dental floss holder, US Patent Application 20 160 008 114, January 14, 2016.
14. J. Wang, A. Olah, and E. Baer, *Polymer*, Vol. 82, p. 166, 2016.
15. S.V. Robinson, Bristled dental floss, US Patent Application 20 170 071 709, March 16, 2017.
16. S. Stuart, Star-shaped microfiber dental floss with nodules, US Patent Application 20 160 242 882, August 25, 2016.
17. Y. David, Dental floss device, US Patent 9 585 737, March 7, 2017.
18. A. Hussain, L. Cahalan, G. Johnson, M. Gartner, P. Cahalan, B.J. Fill, and J.W. Speakman, Surface treated staples, sutures and dental floss and methods of manufacturing the same, US Patent 9 693 841, assigned to Ension, Inc. (Pittsburgh, PA), July 4, 2017.
19. F.J.J. van Velzen, N.P. Lang, E.A.J.M. Schulten, and C.M. ten Bruggenkate, *Clinical Oral Implants Research*, Vol. 27, p. 618, 2016.
20. F.W.M.G. Muniz, H. da Silva Lima, C.K. Rösing, R.S. Martins, M.M.S.M. Moreira, and R.d.S. Carvalho, *Journal of Investigative and Clinical Dentistry*, Vol. in press, p. e12280, 2017.

21. L. Supanitayanon, S. Dechkunakorn, N. Anuwongnukroh, T. Srikhir-in, P. Roongrujimek, and P. Tua-Ngam, Mechanical and physical properties of various types of dental floss, in *Innovative Materials: Engineering and Applications II*, Vol. 730 of *Key Engineering Materials*, pp. 155–160. Trans Tech Publications, 3 2017.
22. J.W. Tukey, *Biometrics*, Vol. 2, p. 99, June 1949.
23. Wikipedia, Tukey's range test — wikipedia, the free encyclopedia, 2017. [Online; accessed 1-December-2017].
24. Wikipedia, Analysis of variance — wikipedia, the free encyclopedia, 2017. [Online; accessed 1-December-2017].
25. H. Bennett and R. Fisher, *The Genetical Theory of Natural Selection*, Oxford University Press, Oxford, 1999.

Index

Acronyms

Chemicals

Boldface numbers refer to Figures

General Index

Also of Interest

Check out these other books by the author published by Scrivener Publishing

Fuel Cells, Solar Panels, and Storage Devices
Materials and Methods
By Johannes Karl Fink
Published 2018. ISBN 978-1-119-48010-5

Chemicals and Methods for Conservation and Restoration
Paintings, Textiles, Fossils, Wood, Stones, Metals, and Glass
By Johannes Karl Fink
Published 2017. ISBN 978-1-119-41824-5

Additives for High Performance Applications
Chemistry and Applications
By Johannes Karl Fink
Published 2017. ISBN 978-1-119-36361-3

Metallized and Magnetic Polymers
By Johannes Karl Fink
Published 2016. ISBN: 9781119242321

Marine, Waterborne, and Water-Resistant Polymers
Chemistry and Applications
By Johannes Karl Fink
Published 2015. ISBN 978 1-119-018486-7

The Chemistry of Printing Inks and Their Electronics and Medical Applications
By Johannes Karl Fink
Published 2015. ISBN 978-1-119-04130-6

The Chemistry of Bio-based Polymers
By Johannes Karl Fink
Published 2014. ISBN 978-1-118-83725-2

Polymeric Sensors and Actuators
By Johannes Karl Fink
Published 2012. ISBN 978-1-118-41408-8

Handbook of Engineering and Specialty Thermoplastics
Part 1: Polyolefins and Styrenics
By Johannes Karl Fink
Published 2010. ISBN 978-0-470-62483-5

Handbook of Engineering and Specialty Thermoplastics
Part 2: Water Soluble Polymers
By Johannes Karl Fink
Published 2011. ISBN 978-1-118-06275-3

A Concise Introduction to Additives for Thermoplastic Polymers
by Johannes Karl Fink.
Published 2010. ISBN 978-0-470-60955-2

Lightning Source UK Ltd.
Milton Keynes UK
UKHW02n1501180418
321277UK00003B/71/P

9 781119 510314